W0246154

STOCHASTIC AND INTEGRAL GEOMETRY

Edited by

R. V. AMBARTZUMIAN
Armenian Academy of Sciences, Yerevan, U.S.S.R.

Reprinted from
Acta Applicandae Mathematicae, Vol. 9, Nos. 1–2 (1987)

D. Reidel Publishing Company / Dordrecht / Boston

Library of Congress Cataloging-in-Publication Data

Stochastic and integral geometry / edited by R. V. Ambartzumian.
  p.   cm.
"Reprinted from Acta applicandae mathematicae, vol. 9, nos. 1–2 (1987)."
ISBN-13: 978-94-010-8239-6   e-ISBN-13: 978-94-009-3921-9
DOI: 10.1007/978-94-009-3921-9
  1. Stochastic geometry—Congresses. 2. Geometry, Integral—Congresses.
I. Ambartzumian, R. V.
[QA273.5.S7   1987]
519.2—dc 19

                                                                87–21719
                                                                   CIP

Published by D. Reidel Publishing Company,
P.O. Box 17, 3300 AA Dordrecht, Holland.

Sold and distributed in the U.S.A. and Canada
by Kluwer Academic Publishers,
101 Philip Drive, Norwell, MA 02061, U.S.A.

In all other countries, sold and distributed
by Kluwer Academic Publishers Group,
P.O. Box 322, 3300 AH Dordrecht, Holland.

All Rights Reserved

© 1987 by D. Reidel Publishing Company, Dordrecht, Holland
Softcover reprint of the hardcover 1st edition 1987
No part of the material protected by this copyright notice may be reproduced or
utilized in any form or by any means, electronic or mechanical,
including photocopying, recording or by any information storage and
retrieval system, without written permission from the copyright owner

TABLE OF CONTENTS

STOCHASTIC AND INTEGRAL GEOMETRY

Edited by
R. V. AMBARTZUMIAN
(Armenian Academy of Sciences, Yerevan, U.S.S.R.)

Acta Applicandae Mathematicae **9** (1987)

© 1987 *by D. Reidel Publishing Company.*

Introduction

The papers in this issue are retrospective versions of several lectures read at the Second Sevan Symposium on Stochastic and Integral Geometry held in Armenia, U.S.S.R., 1–5 October 1985. In choosing this material from the bulk of the work presented at the symposium, we were primarily guided by the desire to represent the more dynamically developing parts of the subject. Although the criterium of thematical integrity has been secondary, the resulting collection does not seem completely void of this property.

Honouring the tradition to dedicate the symposiums on geometrical stochastics (recall the three Buffon symposiums of the past decade) the participants approved dedication of this event to Wilhelm Blaschke to mark 50 years of publication of his *Vorlesungen über Integralgeometrie*. This issue is still another testimony of the great changes which have occurred in the subject so enthusiastically promoted by Blaschke about 50 years ago. Perhaps it will also indicate to the reader some pending developments.

R. V. AMBARTZUMIAN

Acta Applicandae Mathematicae **9** (1987) 3–27.
© 1987 by D. Reidel Publishing Company.

3

Combinatorial Integral Geometry, Metrics, and Zonoids

R. V. AMBARTZUMIAN
Armenian Academy of Sciences, Yerevan 375019, U.S.S.R. .

(Received: 24 June 1986)

Abstract. This paper attempts to look at the interconnections existing between metrics, convexity, and integral geometry from the point of view of combinatorial integral geometry. Along with general expository material, some new concepts and results are presented, in particular the $\sin^2$-representations of breadth functions, translative versions of mean curvature integral, and the notion of 2-zonoids. The main aim is to apply these new ideas for a better understanding of the nature of zonoids.

AMS subject classification (1980). 52A22.

Key words. Translation invariant measures, breadth function, reconstruction from projections, curvatures, wedges, $\sin^2$-representations, zonoids.

The close interrelation between such notions as metrics, convexity, and integral geometry measures have already been disclosed in the classical works of Minkowski and Blaschke. Yet combinatorial integral geometry, which has been developed recently, puts the whole topic in a new light. This paper is an attempt to look at the field from this new point of view. The paper is explanatory in nature so that an uninitiated reader may use it for a first acquaintanceship with the subject. The results referring to $\sin^2$-representations of convex bodies are now published for the first time. Here, a number of basic problems remain unsolved and to promote interest to these questions is one of our main purposes. In Section 6 we outline the connection with the topic known as 'reconstruction from projections'. This, as well as the study of 'roses of hits' in three dimensions, directly leads to the concept of zonoids. To understand their nature remains one of the outstanding problems in convexity theory. Here a new criterion for zonoidality is presented in terms of $\sin^2$-representations. Another result gained within this framework is the expression of the breadth function in terms of curvatures (a translational generalization of the Minkowski's mean curvature integral). A new notion of k-zonoids arises naturally when we apply a theorem of combinatorial integral geometry to convex bodies in $\mathbf{R}^{2k}$. Because this direction remains largely unexplored, we only mention briefly the case $k = 2$.

1. Combinatorial Integral Geometry

The idea of introducing measures into the space of lines in a plane has already been presented in Buffon's classical *needle problem* [1]. Let us recall its formulation. The plane is ruled by a fixed lattice of parallel lines of a certain unit distance apart. A needle l of length $|l|$ is 'thrown' at random onto the plane. What is the probability p of the event that, in its final position, the needle will be intersected by a line of the lattice?

In an equivalent formulation, the needle and the lattice exchange roles and one assumes that it is now the needle which is fixed in the plane with the lattice being thrown down at random.

Without loss of generality, one may assume that the needle lies within some fixed open disc D of unit diameter. Then in all possible outcomes of the lattice-throwing experiment, the disc is intersected by exactly one line of the lattice if we assume that the case of tangency is 'impossible' (i.e., of probability zero). Since other lines of the lattice now play no role, we may fix our attention on this single line, g_D say, intersecting D and Buffon's original problem is now replaced by the following one: what is the probability p that the random line g_D intersecting D should also intersect the needle? We may refer to this as the *dual* problem to the classical Buffon's needle problem.

It is clear that in Buffon's original problem the solution depends on how the experiment is performed, i.e., on the distribution of the final position of the needle with respect to the lattice. Analogously, in the reformulated version, the result depends on the distribution P of the random line g_D.

In the classical solution to Buffon's problem, it is assumed that the centre of the needle and its orientation are independently and uniformly distributed. This means that the projection of the centre onto a line perpendicular to the lines of the lattice, is uniformly distributed on some segment of unit length, and the angle between the line containing the needle and the lines of the lattice is independent and uniformly distributed on $(0, \pi)$. With these assumptions,

$$p = \frac{2|l|}{\pi}.$$

This example is the earliest instance of the calculation of 'geometrical probability'.

To this result corresponds the following solution of the dual problem: there is a unique distribution P of the random line g_D such that

$$P\{g_D \text{ intersects } l\} = \frac{2|l|}{\pi}$$

for every needle l within D. This distribution P is proportional to the restriction to the set of lines intersecting D of the renowned Euclidean motion's invariant measure in the space of lines in the plane. This direct connection with the classical Buffon problem provides justification for terming (as done in the author's monograph [2]) the sets

   $[l] = $ lines that hit l

as 'Buffon sets'.

The first use of other distributions for g_D was made by Bertrand [3] for the purpose of showing that the motion of 'random secant' admitted several interpretations and, therefore, allowed paradoxes ('Bertrand's paradoxes'). These could be avoided by adopting some principle leading to a unique choice among possible distributions, and it was considered that a 'natural' choice was one respecting the relation of Euclidean congruence, in that geometric events, congruent to each other under Euclidean motions, should have equal probabilities. This point of view is clearly expounded by, for example, Deltheil [4].

However, from a more modern point of view, interest in a general P has many other justifications. We mention only the simple observation that if

   $P\{g_D$ hits $\mathscr{P}\} = 0$ for every point $\mathscr{P} \in \mathbf{R}^2$

then the probability

   $P\{g_D$ intersects the needle $l\}$

considered as a function of the needle $l \subset D$ is, in fact, always a continuous linearly-additive pseudometric on D (see Section 2, below).

Remarkably, combinatorial integral geometry has provided tools for proving that the converse is also true (see [2], chapter 6). This amounts to a combinatorial solution of the fourth of Hilbert's famous problems, a connection indicated by Baddeley in his appendix to [2].

Again, combinatorial integral geometry provides tools for disproving a similar result for metrics in $\mathbf{R}^3$.

In the translations-invariant case, this corresponds to the fact that 'zonoids' (see Section 6 below) do not exhaust the class of convex sets in $\mathbf{R}^3$. We note that the first Fragestellung concerning zonoids was due to Blaschke [5] who also coined the term integral geometry [6].

Yet it was Sylvester [7] who, in 1890, first glimpsed the existence of decompositions which later became the cornerstone of combinatorial integral geometry [2]. Sylvester considered the following problem (the Buffon–Sylvester problem in the terminology of [2]).

In the plane, n needles $l_1, \ldots, l_n$ are fixed in a general position. What is the Euclidean motions' invariant measure of the sets

$$\bigcup_1^n [l_i] \quad \text{and} \quad \bigcap_1^n [l_i].$$

Sylvester's result was that the invariant measures in question "become Diophantine linear functions of the sides of the complete $2n$-gonal figure of which the n pairs of extremities of the needles are the angles". Of course, this was a loose expression of the equations of the type (2.4) below.

Neither in [7] nor later was any practical algorithm proposed for finding the corresponding integers and perhaps this is why the whole problem has been somewhat neglected.

Finally, the problem received a very simple solution, which is given in [8]. Soon it became clear that Sylvester's Diophantine decomposition principle is at the source of a potentially vast and fruitful theory. Its basic facts with various ramifications have, for the first time, been systematically expounded in the already-mentioned monograph [2]. Before passing to the main theme (of which the part referring to $\sin^2$-representations is published for first time) we would like to mention several other aspects of this theory.

The theory in many dimensions has an interesting relation to a question posed by Radon. He considered n points $\mathcal{P}_i \subset \mathbf{R}^d$ in the general position and asked for a number of different partitions of $\{\mathcal{P}_i\}$ induced by hyperplanes.

The answer obtained by many authors [9–11] was that this number equals

$$\sum_{K=0}^{d} \binom{n-1}{K}.$$

In the case where d is even, the above obviously equals

$$\binom{n}{2} + \binom{n}{4} + \cdots + \binom{n}{d}$$

that is, the number of odd-dimensional simplices θ_s, which have points from $\{\mathcal{P}_i\}$ for their vertices.

Let $[\theta]$ be the set of hyperplanes hitting the simplex θ. Clearly for every s

$$[\theta_s] = \cup \ a_r$$

where a_r are sets of equivalent hyperplanes: we call two hyperplanes equivalent if they induce the same partition of $\{\mathcal{P}_i\}$. The last relation ignores bundles of hyerplanes through the points $\mathcal{P}_i$. For any measure m in the space of hyperplanes for which m (any bundle) $= 0$, we have

$$m([\theta_s]) = \sum \delta_{sr} m(a_r).$$

There is the fundamental result presented in [12]: *the matrix $\|\delta_{rs}\|$ is square if d is even, and in this case it can be inverted.* From the resulting expression

$$m(a_r) = \sum c_{rs} m([\theta_s]) \tag{1.1}$$

follows the existence of similar decompositions for *every* element $B \in r\{\mathcal{P}_i\}$, the 'Radon ring' generated by the bounded atoms a_r. An algorithm for calculating c_{rs} for $d = 2, 4$ is given in [9].

The decompositions (1.1) have elegant implications in the geometry of the many-dimensional non-Euclidean elliptic spaces. The approach to these questions is based on the possibility of reinterpreting the facts concerning hyperplanes as facts of geometry in elliptical spaces.

This yields a 'combinatorial' version of the Gauss–Bonnet formula for convex polyhedrons in many-dimensional elliptical spaces (see [13]). It gives the volume in terms of the openings of certain angles associated with the polyhedron (only angles having even-dimensional edges participate).

The decomposition (1.1) does not survive for odd-dimensional Euclidean spaces. The situation with planes in $\mathbf{R}^3$ will be considered in Section 5 below.

2. Measures and Metrics

In most books on integral geometry (including [14]) the following problem is discussed in detail.

Given two nonintersecting convex domains D_1 and D_2 in the plane, find the Euclidean motion invariant measure of the set of lines which separate D_1 from D_2.

The solution attributed to Crofton [14] is that the measure in question equals "the least length of a closed string drawn round D_1 and D_2 and crossing over itself at a point minus the lengths of the perimeters of D_1 and D_2" (see Figure 1).

Let us consider a much simpler limiting version of this result in which we replace D_1 by a point $\mathscr{P}_1$ and D_2 by a line segment $\mathscr{P}_2, \mathscr{P}_3$. Thus, the problem is to find the invariant measure of the set

$$[\mathscr{P}_1 \mid \mathscr{P}_2, \mathscr{P}_3] = \text{lines which separate } \mathscr{P}_1 \text{ from } \mathscr{P}_2 \text{ and } \mathscr{P}_3.$$

Using Crofton's theorem, we readily get

$$\mu([\mathscr{P}_1 \mid \mathscr{P}_2, \mathscr{P}_3]) = |\mathscr{P}_1\mathscr{P}_2| + |\mathscr{P}_1\mathscr{P}_3| - |\mathscr{P}_2\mathscr{P}_3|, \tag{2.1}$$

where $|\cdot|$ stands for the Euclidean distance between the points and μ is the Euclidean motion invariant measure.

In fact there is a modification of (2.1) which is valid for rather general measures in the space of lines.

Let us denote by $[\mathscr{P}_1\mathscr{P}_2]$ the set of lines which separate the points $\mathscr{P}_1$ and $\mathscr{P}_2$.

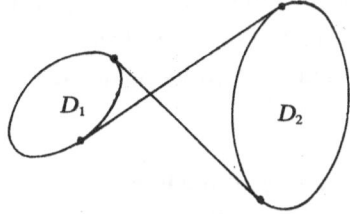

Fig. 1.

Except for the lines through $\mathscr{P}_i$, we have an identity

$$2I_{[\mathscr{P}_1 \mid \mathscr{P}_2 \mathscr{P}_3]}(g) = I_{[\mathscr{P}_1 \mathscr{P}_2]}(g) + I_{[\mathscr{P}_1 \mathscr{P}_3]}(g) - I_{[\mathscr{P}_2 \mathscr{P}_3]}(g),$$

where g denotes a line, I_A is the indicator function of the set A. The proof is by a direct check.

Integration of the above with respect to any measure m in the space of lines which ascribes zero to any bundle of lines through a point yields

$$2m([\mathscr{P}_1 \mid \mathscr{P}_2 \mathscr{P}_3]) = m([\mathscr{P}_1 \mathscr{P}_2]) + m([\mathscr{P}_1 \mathscr{P}_3]) - m([\mathscr{P}_2 \mathscr{P}_3]). \tag{2.2}$$

Clearly, in case $m = \mu$ we have

$$m([\mathscr{P}_1 \mathscr{P}_2]) = 2|\mathscr{P}_1 \mathscr{P}_2|$$

and (2.2) reduces to (2.1). A natural question arises: what are the characteristic properties of the functions

$$\rho(\mathscr{P}_1, \mathscr{P}_2) = m([\mathscr{P}_1, \mathscr{P}_2])? \tag{2.3}$$

If we restrict ourselves to locally-finite measures m whose values on bundles of lines are zero, then it is rather straightforward that *each function ρ is a linearly-additive continuous pseudometric*. In particular, the triangle inequality property follows from (2.2) (where the right-hand side is nonnegative).

Remarkably, a converse statement is also true: *any pseudometric which is linearly-additive and continuous is generated via (2.3) by some measure m in the space of lines from the above-mentioned class and this measure is unique.*

A complete proof of the last statement can be found in [2] where it was derived as a corollary of Diophantine (in the terminology of J. J. Sylvester) decompositions. More properly, we take the planar version of Equation (1.1). Since we have only one-dimensional simplexes (line segments) in the plane, it takes the form

$$m(a_r) = \sum c_{rij} m([\mathscr{P}_i \mathscr{P}_j]). \tag{2.4}$$

If we formally substitute the values $\rho(\mathscr{P}_i, \mathscr{P}_j)$ into (2.4) instead of $m([\mathscr{P}_i, \mathscr{P}_j])$, we obtain a set function defined on the class of sets which can be represented as an a_r for some $\{\mathscr{P}_i\} \subset \mathbf{R}^2$. Then the crucial (and difficult) step is to show that the assumed properties of ρ guarantee that the set function thus defined can be uniquely continued to a measure m in the space of lines which satisfies (2.3).

Pseudometrics, which are invariant with respect to the group $\mathbf{T}_2$ of translations of $\mathbf{R}^2$ ($\mathbf{T}_2$-invariant metrics), are of obvious interest. They are always generated by $\mathbf{T}_2$-invariant measures in the space of lines, and necessarily have the form

$$\rho(\mathscr{P}_1, \mathscr{P}_2) = |\mathscr{P}_1 \mathscr{P}_2| \cdot \lambda(\varphi), \tag{2.5}$$

where the function λ depends solely on the direction φ of the line segment $\mathscr{P}_1 \mathscr{P}_2$. This follows directly from the uniqueness properties of the usual distance on a line.

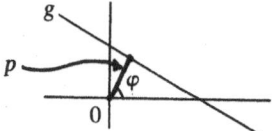

Fig. 2. p is the length of the perpendicular from 0 onto the line g, φ is the direction of g.

On the other hand, any $\mathbf{T}_2$-invariant locally-finite measure in the space of lines has a factorized form

$$dp\nu(d\varphi), \qquad\qquad (2.6)$$

where p, φ are the usual 'polar' coordinates of lines on the plane (see Figure 2) and ν is a symmetrical measure on the circle. (This fact follows easily from the general factorization principles exposed in [15].) It is not difficult to show that *any* choice of totally finite ν in (2.6) produces a locally-finite $\mathbf{T}_2$-invariant measure in the space of lines. Contrary to that, the function $\lambda(\varphi)$ in (2.5) *cannot* be chosen arbitrarily if the aim is to obtain a pseudometric. What is then the appropriate class of functions $\lambda(\varphi)$?

3. Minkowski Metrics

There is an elementary way to answer this question. Let D be a *symmetrical bounded convex domain* in $\mathbf{R}^2$. We define the function $\lambda(\varphi)$ in (2.5) as follows

$$\lambda(\varphi) = (d(\varphi))^{-1}$$

where $d(\varphi)$ is the length of the diameter of D which has direction φ. Let us show that the corresponding ρ is a metric.

As seen from Figure 3.

$$\frac{|0,8|}{|0,7|} = \frac{|0,2|}{|0,1|} \quad \text{and} \quad \frac{|0,6|}{|0,5|} = \frac{|0,3|}{|0,1|},$$

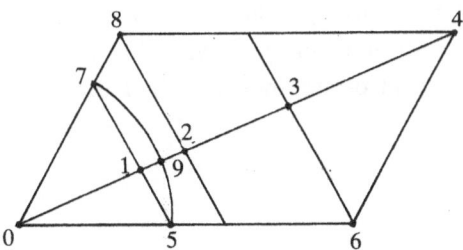

Fig. 3.   Points are denoted by numbers, 0 is the center of symmetry of D. The curve 795 is a part of ∂D. The lines 28 and 36 are parallel to 57.

therefore

$$\frac{|0,8|}{|0,7|}+\frac{|0,6|}{|0,5|}=\frac{|0,2|+|0,3|}{|0,1|}=\frac{|0,4|}{|0,1|}.$$

Since by convexity

$$|0,9| \geqslant |0,1|$$

we will have

$$\frac{|0,8|}{|0,7|}+\frac{|0,6|}{|0,5|} \geqslant \frac{|0,4|}{|0,9|},$$

i.e., the triangle inequality

$$\rho(0,8)+\rho(0,6) \geqslant \rho(0,4).$$

A converse statement is also true. Namely, let ρ be a *metric* satisfying (2.5) (therefore $\lambda(\varphi)$ does not vanish). In the usual polar coordinates (r, φ) on the plane, the equation

$$r=(\lambda(\varphi))^{-1}$$

represents a bounded convex contour. Equivalently, *the unit circle of any* $\mathbf{T}_2$-*invariant metric on* $\mathbf{R}^2$ *is a bounded convex contour.*

Sketch proof: Let $0A$ and $0B$ be two radii of the ρ-circle, i.e.,

$$\rho(0, A) = 1, \qquad \rho(0, B) = 1.$$

Let M be the middle of the segment AB (see Figure 4). It is immediate that the endpoint of the radius in the direction of M lies outside the triangle $0AB$ (otherwise $\rho(0, Q)>2$ and the triangle inequality in $0AQ$ breaks down). This clearly excludes the nonconvex behaviour of the ρ-circle. It remains to consider the case in which a $\mathbf{T}_2$-invariant ρ is a pseudometric in a strict sense, i.e., to assume that

$$\rho(\mathcal{P}_1, \mathcal{P}_2) = 0, \quad \text{for } \mathcal{P}_1 \neq \mathcal{P}_2.$$

Clearly (2.5) implies that $\lambda(\varphi_0) = 0$, where φ_0 is the direction of the segment $\mathcal{P}_1\mathcal{P}_2$. But this can happen only if the corresponding measure ν in (2.6) is proportional to the δ-measure concentrated on $\varphi_0 + \pi/2$. We come to the conclusion that if a $\mathbf{T}_2$-invariant ρ is a strict pseudometric (rather than a metric),

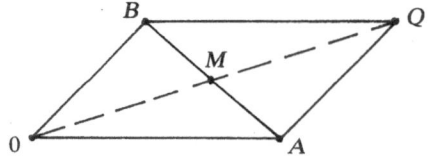

Fig. 4.

then necessarily $\lambda(\varphi)$ is proportional to $|\cos(\varphi - \varphi_0)|$, for some direction φ_0. There is a reformulation of the above criterion which seems to be much wider known. To formulate it we recall the important notion of the *breadth function* $b(\varphi)$ of a convex set. Given convex $D \subset \mathbf{R}^2$, we define $b(\varphi)$ to be the distance between the pair of parallel support lines of D which have common direction φ (see Figure 5; by definition, a support line has a point in common with D but not with the interior of D). In general, $b(\varphi)$ does not determine a convex D in a unique way; but *it does* if we additionally assume that D is centrally symmetrical.

After Minkowski [16], we consider linear continuations b^* of the breadth functions. Given a breadth function $b(\varphi)$ we put

$$b^*(Q) = r \cdot b(\varphi),$$

where (r, φ) are the usual polar coordinates of $\mathscr{P} \in \mathbf{R}^2$, with the origin at the symmetry center of D. There is a proposition due to Minkowski.

PROPOSITION. *Let $b(\varphi)$ be a breadth function of a centrally-symmetrical bounded convex $D \subset \mathbf{R}^2$ with nonempty interior. Then*

$$\rho(\mathscr{P}_1, \mathscr{P}_2) = b^*(\mathscr{P}_2 - \mathscr{P}_1)$$

is a metric.

Conversely, for each $\mathbf{T}_2$-invariant metric ρ its function $\lambda(\varphi)$ is a breadth function of some centrally-symmetric bounded convex D.

The above two propositions imply an interesting property: the equation

$$r = (b(\varphi))^{-1},$$

where $b(\varphi)$ is the breadth function of a centrally-symmetrical domain, in polar coordinates represents the boundary of some other convex domain. Such pairs of domains are called polar reciprocal. Some remarks on this effect can be found in [17].

Both propositions of this section survive in many dimensions (in particular in $\mathbf{R}^3$ on which space we will concentrate below). In fact, similar criteria in $\mathbf{R}^3$ *can be derived* from their planar counterparts. As for the possibility of representation of metrics by measures, we are going to show below that the things in $\mathbf{R}^3$ are substantially more complicated. Often (also in [17]) special terms are used for the

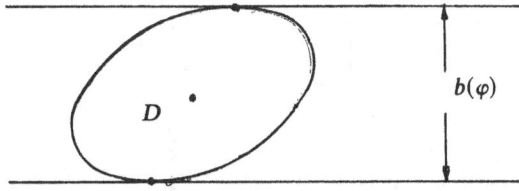

Fig. 5.

basic quantities $\lambda(\varphi)$ and ν, as introduced by (2.5) and (2.6). Namely, given a $\mathbf{T}_2$-invariant measure μ in the space of lines in the plane, the function $\lambda(\varphi)$ is called the *rose of hits* of μ while the measure ν is called *the rose of directions* of μ. This terminology also applies to translation-invariant measures in the space of planes in $\mathbf{R}^3$.

4. Curvature as a Density

By a direct integration of the product measure $d p \nu(d\varphi)$ one easily derives a formula expressing $\lambda(\varphi)$ in terms of ν:

$$\lambda(\varphi) = \int |\cos(\varphi - a)| \nu(da). \tag{4.1}$$

An inverse relation (which gives ν in terms of λ) can be found using considerations from combinatorial integral geometry.

We consider two line segments l_1, l_2 situated as shown in Figure 6. For the rather broad class of measures m, the value $m([l_1] \cap [l_2])$ can be found by the formula

$$2m([l_1] \cap [l_2]) = m([d_1]) + m([d_2]) - m([s_1]) - m([s_2]), \tag{4.2}$$

where as before

$[l_1] =$ the set of lines which hit the segment l_1.

The proof of (4.2) is by summation of the identities written for the sets $[0 \mid BC]$ and $[0 \mid AD]$.

If $m = \mu$, a μ-invariant measure in the space of lines, then

$$\mu([s_1]) = \mu([s_2]) = \lambda(\varphi),$$
$$\mu([d_1]) = \lambda(\varphi + l) \sqrt{1 + l^2} + o(l^2),$$
$$\mu([d_2]) = \lambda(\varphi - l) \sqrt{1 + l^2} + o(l^2),$$

where we assume that the length l tends to zero.

Under appropriate smoothness assumptions on $\lambda(\varphi)$, this yields

$$2\mu([l_1] \cap [l_2]) = [\lambda(\varphi) + \lambda''(\varphi)]l^2 + o(l^2). \tag{4.3}$$

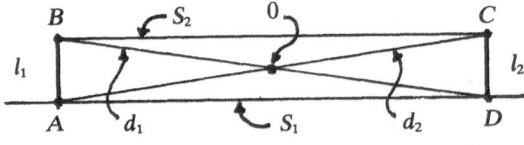

Fig. 6. l_1 and l_2 are perpendicular to s_1 and have a common length l, the direction of s_1 is φ, the Euclidean length of s_1 is one.

Let us note that in the case $\lambda(\varphi) \equiv 2$ where the rose of hits corresponds to the Euclidean motion's invariant measure

$$d\mu^* = dp\,d\varphi$$

the above formula reduces to

$$\mu^*([l_1] \cap [l_2]) = l^2 + o(l^2). \tag{4.4}$$

Let us assume now that our $\mathbf{T}_2$-invariant measure μ has a continuous density

$$d\mu = f(\varphi)\,dp\,d\varphi = f(\varphi) = f(\varphi)\,d\mu^*$$

i.e., $\nu(d\varphi) = f(\varphi)\,d\varphi$. Then $f(\varphi)$ can be calculated as the limit

$$f(\varphi) = \lim_{l \to 0} \mu([l_1] \cap [l_2])(\mu^*([l_1] \cap [l_2]))^{-1}.$$

Using (4.3) and (4.4), we come to the result

$$2f(\varphi) = \lambda(\varphi) + \lambda''(\varphi). \tag{4.5}$$

We compare this with a well-known formula of differential geometry (see [14], page 3)

$$2R(\varphi) = b(\varphi) + b''(\varphi) \tag{4.6}$$

which expresses curvature radius $R(\varphi)$ of a symmetrical convex contour in terms of its breadth function $b(\varphi)$. We conclude that the *density f coincides with the curvature radius* of the convex contour which corresponds to $\lambda(\varphi)$ by Minkowski's proposition of Section 3.

5. Rings in the Space of Planes

In this section we briefly describe the main result of combinatorial integral geometry for *the space of planes* in $\mathbf{R}^3$.

We will denote the space of planes by $\mathbf{E}$.

Let $\mathfrak{W} = \{\mathscr{P}_i\}$ be a finite set of points in $\mathbf{R}^3$ *with no three points on a line*. Each plane $e \in \mathbf{E}$ which does not pass through a point from $\mathfrak{W}$ produces a separation of $\mathfrak{W}$ in two subsets. Two planes e_1, $e_2 \in \mathbf{E}$ are called equivalent if they produce the same separation of $\mathfrak{W}$. Sets of equivalent planes are called *atoms*. All atoms are bounded, except for the one which corresponds to the separation ϕ, $\mathfrak{W}$.

We denote by $r(\mathfrak{W})$ the minimal ring of subsets of $\mathbf{E}$ which contains all bounded atoms (the Radon ring in the terminology of [2]). In this section we formulate the main combinatorial result for general *bundleless* measures m on $\mathbf{E}$, i.e., for measures satisfying the condition

$$m([\mathscr{P}]) \quad \text{for every } \mathscr{P} \in \mathbf{R}^3,$$

where $[\mathscr{P}]$ is the bundle of planes through $\mathscr{P}$. The result is formulated in terms of 'wedges'. In the present context, the term 'wedge' was first introduced in [2], although the corresponding notion has long existed anonymously in integral geometry. A *wedge* is a pair $W = (l, V)$, where l is a finite line segment in $\mathbf{R}^3$, while V is an open domain in $\mathbf{R}^3$ bounded by two planes through l to be termed *faces* of W. V consists of two disjoint parts forming a vertical angle (Figure 7).

A wedge $W = (l, V)$ is said to belong to the *companion* system of the set $\mathfrak{W} = \{\mathscr{P}_i\}$, if

(a) l is of the form $\mathscr{P}_i \mathscr{P}_j$:
(b) the interior of V does not contain any point of $\{\mathscr{P}_i\}$
(c) on each face of W there lies, besides the end-points of l, at least one other point from $\{\mathscr{P}_i\}$.

Thus, the companion system always consists of only finite numbers of wedges.

Given a measure m on $\mathbf{E}$, we define the following 'wedge function' $|W|$:

$$|W| = \frac{1}{2\pi} \int_{[l]} \mathfrak{A}(e)m(de). \tag{5.1}$$

Here $[l]$ denotes the set of planes which hit l, $\mathfrak{A}(e)$ is the opening of the planar angular domain $e \cap V$ (the trace of V on the plane e).

We have the following proposition.

PROPOSITION. *Let m be a bundleless measure on $\mathbf{E}$. For every $B \in r(\mathfrak{W})$ the value $m(B)$ can be calculated as a sum*

$$m(B) = \sum c_s(B)|W_s| \tag{5.2}$$

where $c_s(B)$ are integers which do not depend on the choice of the measure m, the sum is extended over the companion set of wedges.

An algorithm of calculation of the integers $c_s(B)$ can be found in [2], where among other applications, it was also used to find the probability distribution of the number of sides of a random polygon which arises on a random plane sectioning a polyhedron in $\mathbf{R}^3$. Below we consider a rather special case of (5.2) which admits an elementary proof. A general approach to the calculation of the function $|W|$ (in translation-invariant case) will be considered in Section 7.

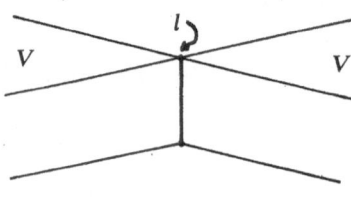

Fig. 7.

Let D be a bounded convex polyhedron in, m be some boundless measure on the space $\mathbf{E}$, and let

$$[D] = \{e \in \mathbf{E} : e \text{ hits } D\}.$$

Clearly $[D]$ belongs to $r(\mathfrak{W})$, where $\mathfrak{W}$ we define to be the set of vertices of D. Therefore, $m([D])$ can be calculated by means of the formulae (5.1)–(5.2).

For this special set, there exists a very simple independent derivation of the same expression. We start from the observation, that for almost every plane which hits D, the intersection $D \cap e$ is a bounded convex polygon whose vertices correspond (in a one-to-one manner) to the edges of D actually hit by e. The elementary fact, that the sum of outer angles of $D \cap e$ equals 2π we write in the form of an identity between indicator functions

$$\sum I_{[a_i]}(e)\, \mathfrak{A}_i(e) = 2\pi I_{[D]}(e). \tag{5.3}$$

Here $[a_i]$ is the set of planes hitting an edge a_i of D, $\mathfrak{A}_i(e)$ is the trace on e of the outer wedge constructed on a_i, and summation is by all edges of D. By definition, for the outer wedge the domain V (see Figure 7) does not possess points in common with the interior of D.

It remains to integrate (5.3) with respect to m:

$$m([D]) = \frac{1}{2\pi} \sum \int_{[a_i]} \mathfrak{A}_i(e)\, m(de)$$

$$= \sum_{\substack{\text{outer wedges} \\ \text{on edges}}} |W_s|. \tag{5.4}$$

In the next section we show that in case where m equals μ, the $\mathbf{M}_3$-invariant measure on $\mathbf{E}$, we have the following wedge function

$$|W| = \tfrac{1}{2}|l| \cdot |V|, \tag{5.5}$$

where $|l|$ denotes the length of l and $|V|$ denotes the opening (flat angle) of the domain V. The results of (5.4) and (5.5) together yield the classical formula

$$\mu([D]) = \tfrac{1}{2} \sum_{\text{edges of } D} |a_i| \cdot |V_i|. \tag{5.6}$$

The wedge function $|W|$ completely determines the parent measure m. This follows from the fact that the class of the sets B for which formula (5.2) applies is sufficiently rich. We also note that if the opening of a wedge W_0 is π, then

$$|W_0| = \tfrac{1}{2} m([l]).$$

Do the values of the function $m([l])$ suffice to completely determine the measure m? In contrast to the planar case, here the answer is known only for the translation-invariant case (see Section 6).

6. Mean Curvature and Funk Problem. Zonoids

Let μ be a locally-finite measure on $\mathbf{E}$ which is invariant with respect to the group $\mathbf{T}_3$ of the translations of $\mathbf{R}^3$ ($\mathbf{T}_3$-invariant). In terms of the parameters p, ω, where

p is the distance of $e \in \mathbf{E}$ from the origin 0, and

ω is the spatial direction normal to e (the orientation of e) usually used to describe planes, such a measure necessarily factorizes:

$$d\mu = dp\nu(d\omega). \tag{6.1}$$

A direct integration of (6.1) yields

$$\rho(\mathscr{P}_1, \mathscr{P}_2) = \mu([\mathscr{P}_1, \mathscr{P}_2]) = \lambda(\xi) \cdot |\mathscr{P}_1\mathscr{P}_2| \tag{6.2}$$

where

$$[\mathscr{P}_1, \mathscr{P}_2] = \{e \in \mathbf{E} : e \text{ separates } \mathscr{P}_1 \text{ from } \mathscr{P}_2\},$$

$|\mathscr{P}_1\mathscr{P}_2|$ is the Euclidean distance between the points,

$$\lambda(\xi) = \int |\cos \widehat{\xi\omega}| \nu(d\omega) \tag{6.3}$$

is called 'the rose of hits' of μ ($\widehat{\xi\omega}$ is the angle between the spatial directions ξ of the segment $\mathscr{P}_1\mathscr{P}_2$ and ω).

The function $\rho(\mathscr{P}_1, \mathscr{P}_2)$ is a pseudo-metric (or a metric if the 'rose of directions' ν is not a δ-measure). A proof follows from the planar result of Section 2 since the image of μ under the map

$$e \to \text{the line } e \cap e_0, \tag{6.4}$$

where e_0 is some fixed plane in $\mathbf{R}^3$, is a measure in the space of lines on e_0.

The question: *does every $\mathbf{T}_3$-invariant continuous, linearly-additive metric ρ in $\mathbf{R}^3$ permit a representation* (6.2) *with some measure μ on $\mathbf{E}$?* has a *negative* answer which we now explain.

Let us fix a plane e_0 and denote by $\mathbf{G}(e_0)$ the space of lines on e_0. Let μ_0 be the measure on $\mathbf{G}(e_0)$ which is the image of μ under (6.4).

For simplicity we assume the existence of a density

$$d\mu = f(\omega) \, d\omega \, dp.$$

Using a standard formula of integral geometry

$$d\omega \, dp = \sin^2 \psi \, d\psi \, d\varphi \, dp'$$

which gives the expression of the Euclidean motions invariant measure in $\mathbf{E}$ in terms of coordinates

φ, p' – parameters determining a line in e_0 (same as in Figure 2).

ψ – the angle between ω and the direction normal to e_0, we find

$$d\mu_0 = F(\varphi)\, d\varphi\, dp \tag{6.5}$$

with

$$F(\varphi) = F(\xi) = \int_{\langle\xi\rangle} f \sin^2\psi\, d\psi. \tag{6.6}$$

We clarify that the direction φ on e_0 has a natural reinterpretation as a direction ξ in $\mathbf{R}^3$ and in (6.6) integration is over the great semicircle $\langle\xi\rangle$ of directions orthogonal to ξ. The condition

$$\int_{\langle\xi\rangle} f \sin^2\psi\, d\psi > 0 \quad \text{for every } e_0 \tag{6.7}$$

suffices to have a continuous, linearly additive metric in $\mathbf{R}^3$ which on *every* plane e_0 corresponds to the measure on $\mathbf{G}(e_0)$ given by (6.5). Yet the condition (6.7) can be met by smooth functions f which are not everywhere nonnegative. For instance, let us take $f_0 = 1$ on the whole hemisphere with the exception of a circle of radius ϵ; in the center of this circle, say, let $f_0 = -1$ and let $|f_0| \leqslant 1$ elsewhere in the circle.

Clearly, under appropriate small choice of ϵ any such (smooth) f_0 will satisfy (6.7). Suppose we insert $f = f_0$ in (6.6). Then the corresponding measure μ_0 will define a metric $\rho_{e_0}(\mathscr{P}_1, \mathscr{P}_2)$ in the plane e_0. Clearly

$$\rho_{e_0}(\mathscr{P}_1, \mathscr{P}_2) = \int_{[\mathscr{P}_1, \mathscr{P}_2]} f_0\, d\omega\, dp$$

and this value does not depend on the choice of e_0 through $\mathscr{P}_1$, $\mathscr{P}_2$. Thus

$$\rho_{e_0}(\mathscr{P}_1, \mathscr{P}_2) = \rho_0(\mathscr{P}_1, \mathscr{P}_2)$$

will be a metric in $\mathbf{R}^3$.

Is this metric ρ_0 in fact generated by some measure μ in $\mathbf{E}$ by means of (6.2)? Because of the uniqueness result in the theory of equations of the (6.3) type (see [18]), the answer to this question is 'no'. In the class of measures μ which possess a sufficiently smooth density, a negative answer follows from the uniqueness of the solution of the so-called *Funk problem*. The latter arises in the present context in the following way.

By applying Equation (4.3) to the measure (6.5) in the space of lines belonging to e_0, we express the integral (6.6) in terms of the rose of hits $\lambda(\xi)$ as

$$\lambda''(\xi) + \lambda(\xi) = 2 \int_{\langle\xi\rangle} f \cdot \sin^2\psi\, d\psi$$

(differentiation is over directions which belong to e_0). Clearly averaging this equation with respect to all possible planes which pass through ξ, yields

$$\Delta_2\lambda(\xi)+2\lambda(\xi) = 2\int_{\langle\xi\rangle} f\,\mathrm{d}\psi \tag{6.8}$$

where Δ_2 is the Laplace operator.

A natural question arises. Suppose we know the rose of hits $\lambda(\xi)$ (and, therefore, the left-hand side of (6.8)). Can we find f, i.e., can we solve the integral equation (6.8) and thus reconstruct the rose of directions? We have come to the classical problem first posed by Funk in [19].

We know the integrals of f over all great circles. Can we reconstruct the function? In a class of smooth functions this problem has a unique solution. Funk, in [19], has shown this by reducing the problem to the classical Abel integral equation.

We may compare (6.8) with the well-known formula of differential geometry, namely (see [5])

$$2(R_1 + R_2) = \Delta_2 b + 2b,$$

where $b = b(\omega)$ is the breadth function of a symmetrical convex body in $\mathbf{R}^3$, R_1, R_2 are the principal radii of the curvature of its surface. We find that in cases where μ has a density f

$$b(\xi) = \lambda(\xi), \tag{6.9}$$

$$R_1 + R_2 = \int_{\langle\xi\rangle} f\,\mathrm{d}\psi. \tag{6.10}$$

It can be shown that the existence of the density f is not necessary for the validity of (6.9). This confirms Minkowski's observation (mentioned in Section 3) that for any measure μ, as in (6.1), its rose of hits coincides with the breadth function of some bounded symmetrical convex body. For the latter we have

$$b(\xi) = \int |\cos\widehat{\xi\omega}|\,\nu(\mathrm{d}\omega), \tag{6.11}$$

The above example of the metric ρ_0 shows that the class of bounded symmetrical convex bodies in $\mathbf{R}^3$ is essentially broader then the class of such bodies whose breadth function admits the representation (6.11) by means of some nonnegative measure ν. The convex bodies that admit (6.11) are called *zonoids* [18] or *Steiner compacts* [20]. We note that a similar notion in the plane is redundant since the breadth function of any planar bounded centrally-symmetrical convex domain can be represented in the form of (4.1) with some totally-finite measure ν on the circle. In the smooth case, this is a consequence of the fact that $R(\varphi)$, as given by (4.6), can always be reinterpreted as the density of a measure.

To give a description of zonoids in terms of differential geometry remains an outstanding problem. This problem is important because the class of zonoidal breadth functions is exactly the domain of solvability of (6.3) viewed as an equation from which the nonnegative measure ν should be determined. To stress the importance of this equation, we briefly describe how it arises in the theory of 'reconstruction from projections'.

Let Γ_β be the space of lines in $\mathbf{R}^3$ which are parallel to a given spatial direction β. Each line $\gamma \in \Gamma_\beta$ can be identified with a point (the trace of γ) on the plane through 0 perpendicular to β. By $d\gamma$ we denote the measure in Γ_β which coincides with the planar Lebesgue measure on the mentioned plane. Let a collection $\{fl_i\}$ of 'flats' be given in $\mathbf{R}^3$: each fl_i (a flat) is a bounded part of a plane from $\mathbf{R}^3$, $\mathfrak{A}_i$ is the direction normal to fl_i, $\|fl_i\|$ is the area of fl_i and $[fl_i]$ is the set of lines in $\mathbf{R}^3$ which hit fl_i.

Clearly, for each i we have

$$\int I_{[fl_i]}(\gamma)\, d\gamma = |\cos \widehat{\beta \mathfrak{A}_i}| \cdot \|fl_i\|$$

(I stands for the indicator function). Summation over i yields

$$\int n(\gamma)\, d\gamma = \sum |\cos \widehat{\beta \mathfrak{A}_i}| \cdot \|fl_i\|$$

$$= A \cdot \sum |\cos \widehat{\beta \mathfrak{A}_i}| \cdot p_i. \qquad (6.12)$$

Here

$$n(\gamma) = \sum I_{[fl_i]}(\gamma)$$

is the number of hits of the line γ with the flats of the collection, A is the sum of the areas of the flats.

$p_i = \|fl_i\| \cdot A^{-1}$ is the probability that a random point dropped on the union of the flats (with uniform distribution) will lie on a flat with orientation $\mathfrak{A}_i$. The quantity $\int n(\gamma)\, d\gamma$ can be termed as the 'cumulative projection' of the collection $\{fl_i\}$ on a plane orthogonal to β (every point of the usual projection is counted with multiplicity $n(\gamma)$).

By an appropriate passage to the limit, Equation (6.12) can be generalized in the form

$$\int n(\gamma)\, d\gamma = \mathbf{A} \cdot \int |\cos \widehat{\beta \mathfrak{A}}| \cdot \mathbf{p}(d\mathfrak{A}) \qquad (6.13)$$

valid for rather general surfaces $\mathbf{S}$ in $\mathbf{R}^3$ having a finite area A. Here on the left-hand side stands the cumulative projection of the surface $\mathbf{S}$, $\mathbf{p}(d\mathfrak{A})$ is a probability distribution in the space of directions, namely the distribution of the normal direction at a point dropped at random with uniform distribution on $\mathbf{S}$.

We conclude from (6.13) that basing upon the knowledge of the cumulative projection of $\mathbf{S}$ on planes of all orientations, we may at best hope for a reconstruction of the probability distribution **p** (other information about **S** is lost). But the kernels in (6.13) and (6.11) coincide.

Several indirect criterions of solvability of Equation (6.13) are given in [18]. Below we outline a new necessary and sufficient condition based on the concept of $\sin^2$-representations.

7. The Wedge Function in the Shift-Invariant Case

How can the function $|W\|$ given by (5.1) actually be calculated? We now give the answer for the $\mathbf{T}_3$-invariant measures on **E** which is especially important because of the connection with the zonoids. Note that because of the factorization (6.1) all $\mathbf{T}_3$-invariant measures are bundleless and (5.2) applies for them without exception.

Now let μ be such a measure. It is convenient to start with the case in which the rose of directions ν is of the delta-type, i.e.,

$$\nu = \delta_\omega \text{ for some spatial direction } \omega.$$

Let $W = (l, V)$ be a wedge (see Figure 7). For a delta-type ν, formula (5.1) yields

$$|W| = F(V) \cdot |l| = \frac{1}{2\pi} |\cos \widehat{\omega l^*}| \cdot \mathfrak{A}(e_\omega) \cdot |l|, \tag{7.1}$$

where $\widehat{\omega l^*}$ is the angle between ω and the direction l^* of the segment l, e_ω is the plane orthogonal to ω.

The formulae of spherical trigonometry enable us to put down an explicit expression for the angle $\mathfrak{A}(e_\omega)$. But we will be interested in the following observation.

Let us assume that l is fixed. Then each domain V (constructed on this l as on Figure 7) can be identified with a pair of symmetrical arcs on a unite circle. The function $F(V)$ defined on such pairs is a measure on $(0, \pi)$ possessing a density $\sin^2 c$.

Here $c = c(\omega, l^*, \phi)$ is the angle between l and the trace of e_ω on the plane $e(\phi)$ (see Figure 8): $e(\phi)$ belongs to the bundle of half planes through l and ϕ determines the angle of its rotation around l.

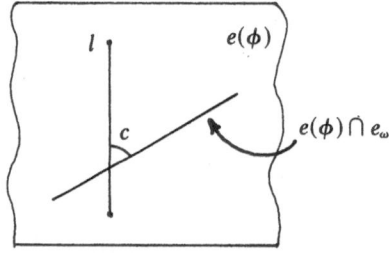

Fig. 8.

Proof. If the opening of V is dv (infinitesimal), then the standard formulae of spherical trigonometry yield

$$\mathfrak{A}(e_\omega) = \frac{\sin c}{\sin \beta} \, dv,$$

where β is the flat angle between e_ω and the plane $e(\phi)$.

The relation

$$\cos \widehat{\omega l^*} = \sin c \sin \beta$$

completes the proof.

It is now straightforward that for a general rose of directions ν, the density of the corresponding F is expressed by the integral

$$f(l^*, \phi) = \frac{1}{2\pi} \int \sin^2 c \, \nu(d\omega). \tag{7.2}$$

Lastly, the wedge function is written in terms of f as follows

$$|W| = |l| \cdot \int_V f(l^*, \phi) \, d\phi. \tag{7.3}$$

Taking into account such symmetry properties as

$$f(l^*, \phi) = f(-l^*, \phi)$$

we can think that the domain of definition of the function f is the product $S_2 \times S_1$ of the unit sphere and the unit circle. It is significant that this space can be interpreted as a group of rotations of $\mathbf{R}^3$ around some fixed point.

EXAMPLE. In case μ is Euclidean motions invariant

$$\nu(d\omega) = d\omega$$

and f reduces to a constant C. Then, according to (7.3)

$$|W| = C \cdot |l| \cdot |V|.$$

To calculate C, we take $|V| = \pi$. By (5.1) $|W|$ will equal half the value of μ on the set of planes hitting l, i.e., $(\pi/2) \cdot |l|$. Hence, (5.5).

8. sin²-Representations of Convex Bodies

The case of wedge density

$$f = \frac{1}{2\pi} \sin^2 c$$

is especially important. We have seen above that this f corresponds to the measure μ_ξ in $\mathbf{E}$ which is a product (compare with (6.1))

$$\mu_\xi(de) = dp \times \delta_\xi(d\omega). \tag{8.1}$$

Let D be a convex domain in $\mathbf{R}^3$ and $[D]$ be the set of planes hitting D. Then a direct integration of (8.1) yields

$$\mu_\xi([D]) = b(\xi), \tag{8.2}$$

the value of the breadth function of D in the direction ξ.

In case D is a polyhedron, formula (5.4), in conjunction with the previous results, yields

$$b(\xi) = (2\pi)^{-1} \cdot \sum_{\substack{\text{outer wedges} \\ \text{of } D}} |a_i| \int_{V_i} \sin^2 c(\xi, a_i^*, \phi) \, d\phi$$

$$= (2\pi)^{-1} \int \sin^2 c \, m_D(d(\Omega, \phi)). \tag{8.3}$$

Here m_D is a symmetrical measure on the space of pairs (Ω, ϕ), i.e., on the product $S_2 \times S_1$ (sphere and circle). This measure is concentrated on the union of the sets

$$\{(a_i^*, \phi) : \phi \in V_i\},$$

is uniform in ϕ on each of these sets and

$$m_D(\{a_i^*, \phi) : \phi \in V_i\}) = |a_i| \cdot |V_i|.$$

Therefore, the total measure

$$m_D(S_2 \times S_1) = \sum |a_i| \cdot |V_i| \tag{8.4}$$

(compare with (5.6)).

The equality (8.3) is our $\sin^2$-representation for polyhedrons. The existence of similar representations for *every* bounded convex body $D \subset \mathbf{R}^3$ now follows by weak convergence arguments. Namely, for any bounded convex $D \subset \mathbf{R}^3$ we can (and do) construct a sequence of uniformly bounded convex polyhedrons D_n such that

$$\lim_{n \to \infty} b_n(\xi) = b(\xi) \quad \text{(pointwise convergence)},$$

where b_n, b are the corresponding breadth functions of D_n, D. In the $\sin^2$-representations (8.3) for $D_n - s$, the total measures of $m_{D_n} - s$ remain bounded (this follows from (8.4)) and, therefore, a weakly convergent subsequence

$$m_{D_{n_i}} \to m$$

can be chosen. Minor complications connected with the fact that the function

$\sin^2 c$ is discontinuous at $\xi = \Omega$ can be easily overcome and we will have

$$b(\xi) = (2\pi)^{-1} \int \sin^2 c(\xi, \Omega, \phi) m(d(\Omega, \phi)) \tag{8.5}$$

a *$\sin^2$-representation for general convex bounded $D \subset \mathbf{R}^3$.*

Given D, the measure m in (8.5) is not uniquely determined (we will see this in Section 10). Yet $\sin^2$-representations may shed new light on some facts of the theory of convex bodies. For instance, Blaschke's basic principle of compactness of sets of uniformly bounded convex bodies now reduces to the similar property of families of measures with respect to weak convergence.

Below we will show that $\sin^2$-representations can be applied to obtain facts previously unknown.

9. $\sin^2$-Representations and Zonoids

Zonotopes are zonoids generated (via (6.11)) by purely atomic measures. This means that the breadth function of a zonoid admits a representation of the type

$$b(\xi) = \sum_1^n |\cos \xi\omega_i| \cdot q_i, \tag{9.1}$$

where ω_i are some directions in $\mathbf{R}^3$ and q_i are some positive weights.

A zonotope is always a bounded convex polyhedron.

A bounded convex polyhedron D is a zonotope if and only if

(a) the collection of edges of D consists of classes such that in each class the edges are parallel and of equal length;
(b) the openings of outer flat angles of the edges sum up to 2π within each class.

The proof that (a), (b) and (9.1) are equivalent is by applying the notion of Minkowski's addition. In fact, (9.1) is a breadth function of Minkowski's sum of n segments whose directions are ω_i and the lengths are q_i. Hence (a), (b).

Conversely, for a polyhedron D, let the conditions (a) and (b) be satisfied. We write the $\sin^2$-representation (8.3) in the form

$$b(\xi) = (2\pi)^{-1} \sum |a_i| \cdot |\cos \widehat{\xi a_i^*}| \cdot \mathfrak{A}_i(e_\xi)$$

$$= (2\pi)^{-1} \sum_{\text{classes}} \sum_{\substack{\text{edges} \\ \text{from a class}}} |a_i| \cdot |\cos \widehat{\xi a_i^*}| \cdot \mathfrak{A}_i(e_\xi)$$

$$= \sum_{\text{classes}} q_j |\cos \widehat{\xi \omega_j}|.$$

where q_j is the common length of the edges in a class and ω_j is their common direction.

From the point of view of $\sin^2$-representations, the above has the following interpretation.

We have a set relation

$$\bigcup_{\substack{\text{edges} \\ \text{from a class}}} \{(a^*_i, \phi) : \phi \in V_i\} = a^*_i \times S_1.$$

It follows from the remarks in Section 8 that on each of the above sets the measure m_D is q_j times the uniform measure on S_1. We come to the conclusion that any zonotope D admits $\sin^2$-representation by means of a *factorized measure on $S_1 \times S_2$ with a uniform factor on S_1*.

What about $\sin^2$-representations for general zonoids?

It is well known that a breadth function of any zonoid admits a pointwise approximation by the breadth function of zonotopes. This means that the measure m in a $\sin^2$-representation (8.5) of a zonoid can be obtained as a weak limit of measures of the mentioned factorized type. The limit will necessarily be of the same factorized type.

Conversely, if the measure m in (8.5) is of the mentioned special type, then a direct integration reduces (8.5) to (6.3).

Hence the following criterium:

a bounded convex $D \subset \mathbf{R}^3$ is a zonoid if and only if its breadth function $b(\xi)$ admits a $\sin^2$-representation by means of a factorized measure on $S_1 \times S_2$, with a uniform factor on S_1; the factor on S_2 coincides with the measure ν in the representation (6.3) of the zonoid.

10. Canonical $\sin^2$-Representations

Formula (5.4) in conjunction with polyhedronal approximation, can be used to obtain $m([D])$ for convex bodies other than polyhedrons. For $\mathbf{T}_3$-invariant measures μ, the result writes in terms of the corresponding wedge function $F(V)$ as in (7.1) and the wedge density f given by (7.2).

Let, for instance, $D = K$, where K is a cone with a unit radius circular base C and constant generator length s. We label the points on ∂C by the angular parameter β, $0 < \beta \leqslant 2\pi$. The generator joining the apex with β is denoted by s_β.

We can approximate K by pyramids. An approximating pyramid will have two types of outer wedges; we give their 'asymptotical' description:

(i) wedges of the form $(d\beta, V_\beta)$ where the line segment $d\beta$ is infinitesimal and belongs to ∂C, V_β is bounded by the plane through C and the tangent plane containing s_β;

(ii) wedges of the form (s_β, V'_β) where the opening of V'_β asymptotically is $s^{-1}\sqrt{s^2 - 1}\, d\beta$.

In the limit, (5.4) takes the form

$$\mu([K]) = \int_0^{2\pi} F(V_\beta)\,d\beta + \sqrt{s^2-1}\int_0^{2\pi} f(s_\beta^*, t_\beta^*)\,d\beta,$$

where t_β is the tangent plane through s_β and t_β^* stands for the corresponding angle.

We can use this formula for the purpose of inversion of (7.2) in case the measure ν has a continuous density, i.e., if

$$\nu(d\omega) = h(\omega)\,d\omega,$$

namely, when $s \to 1$ (the height of K tends to 0) we have

$$\mu([K]) = h(\omega)\mu_0([K]) + o(\mu_0([K])),$$

where μ_0 is the Euclidean motion's invariant measure on $\mathbf{E}$ and ω is the direction normal to the plane of C. In other words,

$$h(\omega) = \lim_{s \to 1} \mu([K])(\mu_0([K]))^{-1}.$$

Here, we have to substitute expression (10.1). After some straightforward analysis, we get the answer

$$h(\omega) = \int_0^{2\pi}\left[2f(\omega, \varphi) - 4\frac{\partial^2}{\partial\phi^2}f(\omega, \varphi)\right]d\varphi. \tag{10.1}$$

Note that here we use the *dual* representation (ω, φ) of a pair (Ω, ϕ), the latter variables have been explained in Section 8: ω is the direction normal to the plane defined by Ω, ϕ while φ is the corresponding angle of rotation around ω. Below, by L we denote the differential operator

$$Lf = 2f - 4\frac{\partial^2}{\partial\phi^2}f.$$

For every convex D we can, using (10.1), express $\mu([D])$ as follows

$$\mu([D]) = \frac{1}{2}\int_{S_2} b(\omega)h(\omega)\,d\omega = \frac{1}{2}\int\int Lbf\,d\omega\,d\varphi.$$

To use the Green formula as we did above, we have to assume (and we do) that ∂D is smooth enough; yet we are no longer bound by any restrictions on the measure ν (equivalently, on μ). We can take, in particular, $\nu = \delta_\xi$ (delta-measure concentrated on a direction ξ); and we will get

$$b(\xi) = \frac{1}{4\pi}\int\int Lb\,\sin^2 c(\xi, \omega, \varphi)\,d\omega\,d\varphi.$$

Here integration in d proves possible. The result will be

$$b(\xi) = \frac{1}{2\pi} \int_{S_2} (R_1 \sin^2 c_I + R_2 \sin^2 c_2) \, d\omega, \tag{10.2}$$

where c_i is the angle between the main direction ϵ_i corresponding to the curvature radius R_i and the trace of e_ξ on the tangent plane with normal direction ω. This is a $\sin^2$-representation in which the measure m on $S_2 \times S_1$ equals

$$d\omega \, x(R_1 \delta_{\epsilon_1}(d\varphi) + R_2 \delta_{\epsilon_2}(d\varphi)).$$

In a sense, (10.2) is a *canonical* $\sin^2$-representation valid for smooth convex bodies. Using other methods, the same result was obtained by the collaborators of the author in [21] and [22].

As a check, the reader may calculate $\int b(\xi) \, d\xi$ by integrating the right-hand side of (10.2). He will get the familiar mean curvature integral. Thus, (10.2) can be considered as a translational counterpart of the latter.

11. Remarks on Convex Bodies in $\mathbf{R}^4$

The results of combinatorial integral geometry in the style of Equation (1.1) can be important in the study of convex bodies in higher dimensions. We restrict ourselves to a corollary of the equation presented on page 77 of [2] which puts the measure on the set $[D]$ of hyperplanes hitting a convex polyhedron $D \subset \mathbf{R}^4$ in the form

$$m([D]) = \tfrac{1}{2} \sum m([\theta_r]) - \tfrac{1}{4} \sum_{\substack{\text{edges} \\ \text{of } D}} m([l_s]). \tag{11.1}$$

Here θ_r are the tetrahedral faces of D and it is assumed that no five vertices of D lie in a hyperplane.

We can again take m to be a product of a δ-measure on S_3 (the three-dimensional sphere) and of one-dimensional Lebesgue measure in the space of perpendicular shifts. Then (10.5) will reduce to an equation between support functions.

We call a convex body in $\mathbf{R}^4$ 2-*zonotope* if its support function can be represented as a sum of the support functions of tetrahedrons in $\mathbf{R}^4$. We call Hausdorff limits of 2-zonotopes 2-*zonoids*. The following theorem is a direct corollary of (10.5):

THEOREM. *The support function of every convex bounded polyhedron* $D \subset \mathbf{R}^4$ *can be represented as a difference of the support function of a* 2-*zonotope and of a zonotope.*

A passage to similar representations for nonpolyhedronal convex bodies requires the calculation of the corresponding limit on the right-hand side of (11.1). Here we have a substantial complication due to the fact that both sums in (11.1) can grow infinitely. In such cases the analysis should involve the appropriate asymptotic decompositions of the sums in question. The infinite terms necessarily annihilate and the nature of the resulting representation will be determined by further terms which are $O(1)$. It seems that in this way one can extend the last statement and obtain a representation result for at least sufficiently smooth convex bodies in $\mathbf{R}^4$.

A justification for such an extension is by its potential usefulness in the study of zonoids in $\mathbf{R}^4$. Indeed, under some symmetry-type conditions, the 2-zonoids happen to be the usual zonoids. Work along these lines is now in progress in the Erevan integral geometry group and the results will be published in due course.

References

1. Buffon, G. L. L.: *Essai d'Arithmetique morale*, Supplement a l'Histoire Naturelle, v. 4, Paris, 1977.
2. Ambartzumian, R. V.: *Combinatorial Integral Geometry*, Wiley, Chichester, New York, 1982.
3. Kendall, M. G. and Moran, P. A. P.: *Geometrical Probability*, Griffin, London, 1963.
4. Deltheil, R.: *Probabilities Geometriques*. Gauthier-Villars, Paris, 1926.
5. Blaschke, W.: *Kreis und Kugel*, de Gruyter, Berlin, 1956.
6. Blaschke, W.: *Vorlesungen über Integralgeometrie I, II*, Hamburger mathematische Einzelschriften, Teubner, Leipzig, 1936, 1937.
7. Sylvester, J. J.: 'On a Funicular Solution of Buffon's "Problem of the Needle" in its Most General Form', *Acta Math.* **14** (1890), 185–205.
8. Ambartzumian, R. V.: 'The Solution to the Buffon–Sylvester Problem in R^3', *Z. Wahrscheinlichkeitstheorie verw. Geb.* **27** (1974), 53–74.
9. Harding, E. F.: 'The Number of Partitions of a Set of N Points in K Dimensions Induced by Hyperplanes', *Proc. Edinburgh Math. Soc. II*, **15** (1967), 285–289.
10. Watson, D.: 'On Partitions of N Points', *Proc. Edinburgh Math. Soc. II*, **16** (1969), 263–264.
11. Schläfli, L.: *Theorie der vielfachen Kontinuität*, Bern, 1952.
12. Ambartzumian, R. V.: 'A Synopsis of Combinatorial Integral Geometry', *Adv. Math.* **37** (1980), 1–15.
13. Ambartzumian, R. V.: 'On Combinatorial Foundations of Integral Geometry', *Izvestia Acad. Sci. Armenian SSR, Mathematics* (English: *Soviet J. Contemporary Analysis*) **16**, 4 (1981), 285–291.
14. Santalo, L. A.: *Integral Geometry and Geometric Probability*, Addison-Wesley, Reading, Mass., 1976.
15. Ambartzumian, R. V.: 'Factorization in Integral and Stochastic Geometry', *Teubner-Texte zur Math.* v. 65, 1984, pp. 14–33.
16. Minkowski, H.: 'Volumen und Oberfläche', *Math. Ann.* **57** (1903), 447–495.
17. Stoyan, D. and Mecke, J.: *Stochastische Geometrie*, Akademie-Verlag, Berlin, 1983.
18. Schneider, R. and Weil, W.: 'Zonoids and Related Topics', in *Convexity and its Applications*, Birkhäuser, Basle, 1983, pp. 296–317.
19. Funk, P.: 'Über Flächen mit lauter geschlossenen geodätischen Linien', *Math. Ann.* **74** (1913), 278–300.
20. Matheron, G.: *Random Sets and Integral Geometry*, Wiley, New York, 1975.
21. Panina, G. J.: 'Translation-invariant Measures and Convex Bodies in R^3', *Zapiski nauchnih seminarov LOMI* **157**, 1987.
22. Aramian, R. G.: 'On Stochastic Approximation of Convex Bodies', in print.

Acta Applicandae Mathematicae **9** (1987), 29–60.
© 1987 *by D. Reidel Publishing Company.*

Stochastic Differential Geometry:
An Introduction

WILFRID S. KENDALL
*Department of Mathematics, Strathclyde University, Livingstone Tower, 26 Richmond Street,
Glasgow G1 1XH, Scotland*

(Received: 16 January 1986)

Abstract. Stochastic calculus can be used to provide a satisfactory theory of random processes on differentiable manifolds and, in particular, a description of Brownian motion on a Riemannian manifold which lends itself to constructions generalizing the classical development of smooth paths on a manifold. An introduction to this theory is given, and a survey is made of the relationship between curvature properties of the manifold and the asymptotic behaviour of the Brownian motion on the manifold. It is then explained how these results can be used to prove geometrical theorems concerning special classes of maps between manifolds.

AHS subject classifications (1980). 58G32, 60J65, 60H10.

Key words. Semimartingale, Γ-martingale, Brownian motion on a manifold, geodesic, connection, parallel transport, stochastic lift, sectional curvature, Ricci curvature, harmonic map.

0. Introduction

A growing theme of modern probability is the resurgence of geometry. On the applied side this shows most clearly in the topics of stochastic geometry, and integral geometry as applied to stereology, as described elsewhere in this volume. It is displayed in the theoretical aspect of probability by the study of random processes on nonlinear state-spaces or *manifolds*, such as spheres, rotation groups, and projective spaces, using the stochastic calculus. This paper is intended to introduce the study of these random processes. I have tried to write it in such a way as to offer some insight both to probabilists and to geometers, having met members of both classes of mathematician who expressed desires to learn more about this meeting-ground between the two disciplines. This objective has enforced many compromises; one cannot possibly give proper introductions to both differential geometry and stochastic calculus in the same paper. Consequently geometry and probability are discussed in summary fashion; this is not a detailed map of the subject but rather a preliminary guide.

For the probabilist, the study of random processes on manifolds is an obvious sequel to that of random processes on Euclidean spaces. Early examples include work by Itô (1950), Dynkin (1961), and Yosida (1949). Recent studies have clarified the concepts, allowing a more direct engagement with the actual paths of the random processes and a clarity of expression arising from coordinate-free definitions.

Although stochastic differential geometry has not yet made great inroads on applied probability, such inroads await only time and greater familiarity with the subject. As indications of possibilities we quote D. G. Kendall's study of the diffusion of shape (D. G. Kendall, 1977), the work of Antonelli and others on diffusions and manifolds in genetics (see Antonelli *et al.*, 1980, and references therein), and the intriguing study of random rotations arising in physics (see, for example, the book of McConnell, 1980).

But stochastic differential geometry need not wait on applied probability in order to find applications. It is already the case that stochastic differential geometry has much to offer nonstochastic differential geometry, primarily because of the well-known relationship between diffusions and second-order elliptic differential operators. The Laplace–Beltrami operator is a differential operator on a Riemannian manifold that generalizes the Euclidean Laplacian. It is intimately connected with Brownian motion on a manifold; one way of understanding this connection is explained in Sections 1 to 3 of this paper. The Laplace–Beltrami operator itself reflects much of the geometry of a manifold (see the recent book by Chavel, 1984); the probabilist can understand this in a particularly direct way by means of analysis of suitable stochastic differential equations. Sections 4 and 5 provide an initiation into this far-reaching procedure.

However, the use of stochastic differential geometry is not confined to the study of Laplace–Beltrami operators. As described in Section 6, more general processes than Brownian motion can be used to prove theorems about the important class of *harmonic* maps between manifolds. As well as providing a probabilistic approach to the theory of harmonic maps, these results have suggested new ideas in stochastic differential geometry which are worthy of investigation in their own right.

As might be expected, stochastic calculus is a prerequisite for a proper understanding of stochastic differential geometry, being required both in order to work with differentials of the random paths, and also in order to define geometrically natural processes in ways that do not depend on particular coordinate systems. Sections 1 and 2 give a brief summary of stochastic calculus; the novice will find many monographs on the subject, of which we mention Chung and Williams (1984) as an introduction, and Dellacherie and Meyer (1978, 1982), Ikeda and Watanabe (1981), and McKean (1969) as more extensive treatments.

Of course the stochastic differential geometer must also be familiar with nonstochastic differential geometry, which provides much of the conceptual world of the subject. Some description of the most important geometrical concepts (but avoiding technicalities) is to be found, in particular, in Sections 2, 3, and 4. Much of what differential geometry has to offer lies in clear understanding of these technicalities; for this we refer the reader to the textbooks mentioned in the course of the paper and in particular to those by Bishop and

Crittenden (1964), Cheeger and Ebin (1975), Chavel (1984), the appendix of Elworthy (1982), and Warner (1971).

The treatment of stochastic differential geometry in this paper is naturally biased towards the author's own interests. I make no apology for this; length considerations prohibit a complete coverage. Section 7 is a token survey; a series of brief comments on other branches of stochastic differential geometry (interpreting the phrase in a broad sense). To keep length under control, only a representative set of references have been given. There are several surveys and books slanted towards other aspects of the subject: Pinsky (1978, 1983), Molchanov (1975), Elworthy (1982), Ikeda and Watanabe (1981), Bismut (1981), Malliavin (1978).

1. Real-Valued Semimartingales

At the heart of nonstochastic differential geometry lies the study of curves defined by geometrically natural differential equations. Correspondingly *stochastic* differential geometry is built on the analysis of random continuous curves and therefore requires a *stochastic* calculus. In general, the random curve need no longer be differentiable, but must possess a stochastic regularity or *semimartingale property*. Differential equations are replaced in stochastic differential geometry by equations involving *stochastic differentials*, for which there is a well-developed theory of existence and uniqueness. The purposes of stochastic differential geometry are best served if these equations can somehow be expressed in an invariant manner, so that the corresponding solutions do not depend on the particular choice of coordinate system. In this section the basic theory of stochastic calculus is surveyed briefly, as a preliminary to the study of stochastic differential geometry.

The fundamental notion is that of a (continuous) semimartingale. A *real-valued semimartingale* is a random continuous path on the real line

$$X: [0, T) \to \mathbf{R}$$

which can be expressed as a sum of 'noise' plus 'systematic component' $X = M + V$ in a fashion that is essentially unique. The second component V is of locally bounded variation and so satisfies a generalized classical differential equation with random coefficients; it is the systematic component of the random path. The first component M is the *martingale* or noise component of the random path; it contains no systematic component of variation (in a sense made precise below). The terminal time T may be infinite, or may be a random *stopping time* as defined below.

To ensure uniqueness of the decomposition we set $M(0) = 0$ in the sequel.

The precise requirement on V is for it to be a continuous *predictable* random process of locally bounded variation. In the cases discussed here predictability corresponds to V being *adapted*; for all times t the random variable $V(t)$ is measurable with respect to $\mathcal{F}_t$ the σ-field of events depending only on information prior to time t. The systematic nature of V can be appreciated by noting that it is an infinitesimal analogue of a one-step-look-ahead prediction of X:

$$E(X_{t+\delta t} \mid \mathcal{F}_t) - X_t \simeq \delta V. \tag{1.1}$$

Since V is of locally bounded variation, it is susceptible to the methods of ordinary differential calculus.

The other term $M = X - V$ is a continuous *local martingale*. It is a *martingale* if

$$E(M_{t+s} \mid \mathcal{F}_t) = M_t \quad \text{for all nonnegative } s \text{ and } t \tag{1.2}$$

(the most important examples for us are the famous Brownian motion or *Wiener process* W, and stochastic integrals built on W). A local martingale M is a continuous random process for which there exists an increasing sequence $\{T_n\}$ of stopping times tending to T such that $\{X(t \wedge T_n) : t \geq 0\}$ is a continuous martingale for each n. A *stopping time* T_n is a nonnegative random variable for which

$$\{T_n \leq t\} \in \mathcal{F}_t \quad \text{for all } t. \tag{1.3}$$

Thus, the event that T_n occurs before t depends on events that are determined by time t.

A semimartingale on Euclidean space is a random process each of whose coordinate processes is a real-valued semimartingale. Because the martingale property is linear, it does not matter here which set of linear coordinates are taken for Euclidean space. However, stochastic differential geometry requires consideration of random processes on essentially nonlinear *manifolds*. The martingale property is not preserved by nonlinear transformations. Nevertheless, stochastic calculus can justify an extension of the semimartingale property to manifold-valued processes.

Before leaving this topic we emphasize that all semimartingales considered in this paper are *continuous* random functions of time. There is an extensive theory for *discontinuous* semimartingales (see Dellacherie and Meyer; 1978, 1982, for an authoritative account) but this has not yet found much application in stochastic differential geometry.

1.1. STOCHASTIC CALCULUS AND ITÔ'S LEMMA

The fundamental observation underlying stochastic calculus is that the process $f(X)$ is a semimartingale whenever f is twice-differentiable and X is a semimartingale. Because of this, it is meaningful to speak of semimartingales taking values in C^2-manifolds (nonlinear spaces **M** for which there is a notion of a twice-differentiable function). A process X on such an **M** is a semimartingale if and only if $f(X)$ is a real-valued semimartingale for all f in $C^2(\mathbf{M})$.

In such a case $f(X)$ has a unique decomposition as local martingale plus adapted process of locally bounded variation. If X is a real-valued semimartingale, then Itô's lemma expresses $f(X)$ in terms of $X = M + V$.

THEOREM 1. *Itô's Lemma.*

$$f(X_t) - f(X_0) = \int f'(X)\, d_I X + \tfrac{1}{2} \int f''(X)\, d[X, X] \qquad (1.4)$$

where

$$\int f'(X)\, d_I X = \int f'(X)\, d_I M + \int f'(X)\, dV.$$

The proof of Itô's lemma depends on analysis of the second-order Taylor expansion of $f(X)$. The integrals are those of stochastic calculus and we discuss them briefly now.

The integrator $[X, X]$ in the second integral is the locally-bounded-variation component of the semimartingale M^2 and is referred to as the *bracket process* of $X = M + V$. (The bracket process $[X, Y]$ of two semimartingales is the locally-bounded-variation component of the product of the corresponding two local martingales. It measures a kind of infinitesimal variance and covariance for semimartingales.) In particular, the martingale M is constant on intervals of constancy of $[M, M]$, and $[M, M]$ is approximated by the quadratic variation

$$\sum_k (M((k+1)2^{-n}) - M(k2^{-n}))^2 \qquad (1.5)$$

as n tends to ∞.

The integral $\int f'(X)\, d_I M$ is the famous Itô integral expounded, for example, in McKean (1969) in the case of the Wiener process, and in general in Ikeda and Watanabe (1981). The suffix I in d_I warns against the interpretation of $\int f'(X)\, d_I M$ as an ordinary integral. More generally, $\int F\, d_I M$ is defined for all F which are (a) predictable (continuous and adapted will suffice in all cases we consider) and (b) satisfy

$$\int F^2\, d[M, M] < \infty. \qquad (1.6)$$

The integral $\int F\, d_I M$ is linear in both F and M and is constant on intervals of constancy of either F or M.

It will be convenient to abbreviate equations involving Itô integrals by using stochastic differentials as below

$$d_I f(X) = f'(X)\, d_I X + \tfrac{1}{2} f''(X)\, d[X, X] \qquad (1.7)$$

when f is C^2. To interpret such an equation one need only multiply both sides by a suitable F and integrate.

1.2. REAL-VALUED BROWNIAN MOTION

A continuous real-valued process W is a Brownian motion (we write $W \sim$ BM(R)) if:

it begins at zero,

if $W(t + s) - W(t)$ is normal of mean zero and variance s,

and if $W(t + s) - W(t)$ is independent of the past $\mathcal{F}_t$ for all positive t and s.

Some simple facts about BM(R) are basic tools for stochastic differential geometry:

$$\liminf W_t = -\infty \quad \text{as } t \text{ tends to } \infty,$$

$$\limsup W_t = \infty \quad \text{as } t \text{ tends to } \infty, \tag{1.8}$$

$$\lim(W_t + ct)/t = c \quad \text{as } t \text{ tends to } \infty.$$

THEOREM 2 (Lévy). *The process X is* BM(R) *if and only if*

$$X \text{ is a local martingale} \quad and \quad [X, X](t) = t \text{ for all } t. \tag{1.9}$$

This generalizes to higher dimensions, characterizing $X = (X^1, \ldots, X^n)$ as Brownian motion on the Euclidean space R^n (we write $X \sim$ BM(R^n)) if and only if it is a local martingale and

$$[X_i, X_j](t) = t \quad \text{if } i = j, 0 \text{ otherwise.} \tag{1.10}$$

The martingale characterization of Stroock and Varadhan (1979) is a far-reaching generalization of Lévy's theorem.

1.3. STOCHASTIC DIFFERENTIAL EQUATIONS

Semimartingales arise in practice as explicit solutions to stochastic differential equations such as

$$d_t X = f(X) \, d_t W + g(X) \, dt. \tag{1.11}$$

If f and g are uniformly Lipschitz, then there is a unique continuous semimartingale solution for every initial value $X_0 = x_0$. If the Lipschitz conditions are not uniform, then solutions may be defined only up to an explosion time ζ. At this time the solution ceases to be defined; by time ζ it has left all compact sets.

More generally, the Brownian motion W can be replaced by a general continuous semimartingale. The multidimensional generalization is straightforward.

The solution X actually depends smoothly on the initial data for the stochastic differential equation above. This is of importance in the theory of stochastic flows; however we will not investigate it here.

In order to draw conclusions about solutions to stochastic differential equations we require a comparison theorem (as for example in Yamada, 1973).

THEOREM 3 (Comparison of stochastic differential equations). *Suppose Y and Z solve*

$$d_I Y = d_i W + f(Y) \, dt, \qquad d_I Z = d_I W + g(Z) \, dt$$

with $Y_0 > Z_0$ and with f and g uniformly Lipschitz. If $f > g$ then $Y > Z$ for all time.

The proof analyzes the Itô differential of $U = \log(Y - Z)$. The martingale parts of Y and Z coincide so the process U is of locally bounded variation and indeed

$$U_t - U_0 > - \text{constant} \cdot t$$

Hence, $Y - Z$ can never reach zero.

A modified version also holds for locally Lipschitz coefficients. In this case we must allow for the possibility of explosion.

1.4. STRATONOVICH CALCULUS

Stochastic differential geometry specifies particular random processes on manifolds as solutions to naturally arising stochastic differential equations. This requires us to come to terms with the fact that semimartingales on general nonlinear manifolds do not have the well-defined decompositions of above. However, if we employ the *Stratonovich integral*

$$\int Y \, d_S X = \int Y \, d_I X + \tfrac{1}{2} [Y, X] \tag{1.12}$$

(defined when both X and Y are semimartingales) then the correction term means that the Stratonovich differentials satisfy the usual rules of calculus:

$$d_S f(X) = f'(X) \, d_S X \quad \text{if } f \text{ is } C^3. \tag{1.13}$$

Note that an extra degree of differentiability is required of the function f, and that the stochastic differential equation makes sense only if multiplied by a semimartingale (rather than a more general adapted process) before integrating. In return it becomes easy to write down intrinsic differential definitions of important semimartingales, as discussed in the next section. As a general (but not invariable) rule, Stratonovich differential equations are used to provide intrinsic definitions of a semimartingale while Itô calculus (because of its close relationship to martingales) assists in deriving statistical properties.

To close this section we note that the equations

$$d_S Z_1 = Y_1 \, d_S X_1, \qquad d_S Z_2 = Y_2 \, d_S X_2$$

imply

$$d[Z_1, Z_2] = Y_1 Y_2 \, d[X_1, X_2].$$

This is useful in relating Stratonovich differential equations to the Itô calculus.

2. Calculus on Manifolds

The formula $d_S f(X) = f'(X) d_S X$ shows that $d_S X$ behaves as an ordinary differential. Thus, if X is a semimartingale on a manifold, then its Stratonovich differential can be viewed formally as a tangent vector or velocity. In order to investigate this further, we recall briefly the notions of manifolds and tangent vectors; for thorough treatments see, for example, Milnor (1963), Spivak (1965), or Warner (1971).

For our purposes it is sufficient to visualize a manifold $\mathbf{M}$ as a star-shaped open subset U of R^m together with some (possibly very complicated) identifications of the boundary ∂U with itself. Thus, in the case of the sphere S^2 one may obtain U by deleting the north pole, spreading out the remainder of S^2 on the plane in a distorted projection, and identifying all of the boundary ∂U to be one point (the missing north pole).

Such representations depend on arbitrary choices, such as nominating a particular point of S^2 to be the north pole. In the definitions and calculations of stochastic differential geometry, it is useful if definitions and calculations make no reference to a particular coordinatization or representation of the manifold in question; this leads to greater clarity and efficiency of thought and also avoids definitions that accidentally depend on the particular representation chosen.

Tangent vectors can be understood by using a dynamical interpretation. A smooth path $\{x(t): t \geqslant 0\}$ on a smooth manifold $\mathbf{M}$ possesses a velocity $v = (dx/dt)|_{t=0}$ at $x(0)$, and v can be viewed as a vector in an m-dimensional space. Moreover, v will depend only on the first-order behaviour of $x(t)$ at $t = 0$ and will be essentially independent of the coordinatization used to calculate the derivative. However, the vector space depends on the location $x(0)$; velocities attached to different points cannot, in general, be added together in any meaningful sense (think of velocities of particles at different locations on a sphere). Consequently with each velocity record must be kept of the corresponding location x; the set of all pairs (x, v) forms the tangent bundle $T\mathbf{M}$ of $\mathbf{M}$. The tangent bundle $T\mathbf{M}$ itself can be given the structure of a smooth manifold.

We may think of $T\mathbf{M}$ as a product space $U \times R^m$ together with identifications on the boundary $\partial U \times R^m$. However, it should always be borne in mind that if x and y are distinct members of $\mathbf{M}$ then, in general, no meaning can be attached to the sum of two tangent vectors (x, u) and (y, v). It is, however, meaningful to add two velocities attached to the same base-point x. Such tangent vectors comprise the fibre of $T\mathbf{M}$ at x

$$T_x\mathbf{M} = \{(x, v): v \text{ is a velocity at } x\}. \tag{2.1}$$

Given two manifolds $\mathbf{M}$ and $\mathbf{N}$ and a smooth map $F: \mathbf{M} \mapsto \mathbf{N}$, consider a tangent vector (x, v) for $\mathbf{M}$. This will be produced by some path $x(t)$ with $x(0) = x$ and so we can set

$$TF(x) \cdot v = \left(F(x), \frac{d}{dx} F(x(t))|_{t=0} \right). \tag{2.2}$$

Here T stands for 'tangent'. TF is the tangent map or derivative of F. Of course, it is necessary to show that TF does not depend on the particular choice of the curve $x(t)$ representing (x, v).

Because the Stratonovich differential obeys the ordinary rules of calculus, it can be shown that the semimartingale $F(X)$ satisfies

$$d_S F(X) = TF(X) \cdot d_S X \tag{2.3}$$

no matter what coordinate system is chosen. Thus, $d_S X$ can indeed be viewed as a formal velocity. The same cannot, of course, be said for $d_I X$, whose interpretation depends in general on a choice of coordinate system.

For f a smooth real-valued function on $\mathbf{M}$ it is convenient to write

$$\frac{d}{dt} f(x(t)) |_{t=0} = v \cdot f(x) \tag{2.4}$$

and

$$Tf(x) \cdot v = (f(x), v \cdot f(x)) \tag{2.5}$$

where v is the velocity of $x(t)$ at $t = 0$. Thus, the tangent vector (x, v) is regarded as acting on f by differentiating it at x in the direction v.

We consider one more geometrical notion before turning to the theory of semimartingales on manifolds. We have noted that vectors in different fibres of $T\mathbf{M}$ cannot be added together. However, it is often appropriate to impose a notion of length for tangent vectors, by assigning a positive-definite inner product to each fibre in a smooth fashion. Two velocities v and u at a fixed point then have an inner product $\langle u, v \rangle$. This inner product is called a Riemannian metric. When equipped with a Riemannian metric $\mathbf{M}$ is called a Riemannian manifold.

3. Semimartingales on Manifolds

Section 2 explained what it means for a random process X on a smooth manifold $\mathbf{M}$ to be a semimartingale; $f(X)$ must be a semimartingale for all smooth real-valued functions f. For a general manifold the martingale property has no meaning; however, it is still valid to use stochastic calculus employing Stratonovich differentials. The resulting semimartingales will not depend on the particular coordinate representation employed for $\mathbf{M}$. In this section we describe a fundamental system of stochastic differential equations (due to the work of Gangolli, 1964, and Eells and Elworthy, 1970, see also Malliavin, 1978) that relates X to a semimartingale on Euclidean space (its *stochastic development*) and to a further semimartingale that describes a moving coordinate frame of reference for X (its *stochastic parallel transport*). To define these associated processes, it is necessary to suppose that $\mathbf{M}$ is equipped with an extra piece of geometric structure known as a *connection*. If one considers how the associated processes might be defined, then the properties of a connection seem quite natural.

In the sequel we shall take **M** to be a Riemannian manifold, so that each tangent fibre $T_x\mathbf{M}$ has a Euclidean metric. This is not the most general case but contains all the essential features of interest.

3.1. STOCHASTIC DEVELOPMENT AND PARALLEL TRANSPORT

We have discussed above how the Stratonovich differential $d_S X$ is associated in some formal sense to the tangent bundle and in particular to the fibre $T_X\mathbf{M}$ sitting above X. The fibre is isometric to Euclidean space **V** of the same dimension m as **M**; let $O_X(\mathbf{M})$ be the collection of all isometries from **V** to $T_X\mathbf{M}$ and $O(\mathbf{M})$ be the union of all fibres $O_X(\mathbf{M})$ as x runs through **M**, the orthonormal frame bundle. Then $d_S X$ can be 'pulled back' to a Stratonovich differential on **V** if only we can find a semimartingale Ξ with Ξ_t belonging to that fibre of $O(\mathbf{M})$ that sits above X_t. (It is possible to give $O(\mathbf{M})$ the structure of a smooth manifold so it makes sense to speak of Ξ as a semimartingale.) Let π be the obvious projection $\pi\colon O(\mathbf{M}) \to \mathbf{M}$. We require a semimartingale Ξ with $\pi(\Xi) = X$. Given such a parallel transport Ξ, we define the stochastic development Y of X onto **V** by

$$d_S Y = \Xi^{-1} d_S X \tag{3.1}$$

(note that Ξ^{-1} is also a semimartingale on a suitable manifold so that the Stratonovich differential equation above makes sense).

The crux of the matter is to see how to define Ξ. An expression must be found for $d_S \Xi$, which is formally associated to $T_\Xi O(\mathbf{M})$. Since we are only given $d_S X$, which is associated to $T_X\mathbf{M}$, it is clearly necessary to choose for each ξ in $O(\mathbf{M})$ a linear map H_ξ which connects $T_{\pi\xi}\mathbf{M}$ to $T_\xi O(\mathbf{M})$:

$$H_\xi\colon T_{\pi\xi}\mathbf{M} \to T_\xi O(\mathbf{M}).$$

If H itself varies smoothly with ξ then

$$d_S\Xi = H_\Xi\, d_S X \tag{3.2}$$

is a stochastic differential equation with smooth coefficients and so the solution Ξ exists up to a (possibly finite) explosion time.

Of course we require $\pi\Xi = X$ and this holds if and only if

$$(T\pi)\, H_\xi = \text{identity on } T_{\pi\xi}\mathbf{M}$$

for all ξ in $O(\mathbf{M})$. Here $T\pi$ is the tangent map corresponding to the projection π. This condition is easily verified by differentiating the relation $\pi\Xi = X$.

A choice of maps H represents a connection if it satisfies this condition and also a more technical equivariance condition which assures us of the following; though Ξ depends on its initial condition $\Xi_0 = \xi$, yet a change in the initial condition that leaves $\pi\xi$ fixed should only have the effect of subjecting the solution Ξ to a fixed isometrical transformation of the model space **V**. For the sake of brevity, we say no more about the second condition.

The explosion time for the stochastic differential equation

$$d_S\Xi = H_\Xi\, d_S X$$

is actually the same as the explosion time for X (understood as occurring when X explodes out of all compact sets). In the Riemannian case, this follows quickly because the fibres of the orthonormal frame bundle are all compact; Elworthy (1982, p. 175 Thm. 13c) proves the generalization for non-Riemannian manifolds.

It is convenient to write the system of stochastic differential equations as

$$d_S X = \Xi\, d\, Y_S, \qquad d_S\Xi = H_\Xi\, d_S X. \tag{3.3}$$

The stochastic development Y can be thought of as the path traced out on **V** when **V** is rolled without slipping over **M**, always keeping the point of contact on **M** to be X. The parallel transport Ξ describes the orientation of **V** as it rolls and may be thought of as the inertial frame of reference carried about by the semimartingale X.

When X yields a smooth path, the development and transport are exactly those described in the nonstochastic theory of differential geometry. The dominant role played by the connection justifies the description of the theory as stochastic *differential* geometry rather than Riemannian geometry; the choice of the map H is not uniquely determined by the Riemannian metric. However, the metric does have associated with it one connection of special significance, to be discussed shortly.

3.2. CONNECTIONS AND FAMILIES OF SEMIMARTINGALES

The stochastic differential system (3.3) can be used to categorize semimartingales X on **M** in terms of their stochastic developments Y on Euclidean space **V**. We emphasize the essential role of the connection, and the manifold state-space of semimartingales concerned, by using the prefix 'Γ'. (Γ is the symbol used conventionally in coordinate notation for connections.)

We view X as being driven by Y using the system (3.3). Note that, though a driving process Y can be defined for all times, the solution X may explode at a time ζ

$$\zeta = \sup\{t : X \text{ leaves } K \text{ at time } t, \text{ where } K \text{ is compact}\}. \tag{3.4}$$

However, as noted above, the parallel transport Ξ will not explode before X. Note also that if **M** is compact then the explosion time is almost surely infinite.

We will be concerned with four important classes of semimartingale.

DEFINITION (Γ-geodesics). If $Y_t = v \cdot t$ for a fixed constant vector p in $\mathbf{V}$ and $\Xi_0 = \xi$ then X is said to be a Γ-geodesic with initial velocity $\xi \cdot v$.

This definition is basic to nonstochastic differential geometry. It also underlies calculations in stochastic differential geometry as shown in the next section. The equivariance condition on H referred to above means that a Γ-geodesic depends on its initial conditions ξ, v only through its initial tangent vector $\xi \cdot v$.

DEFINITION (Γ-Brownian motion). If Y is BM($\mathbf{V}$) then X is said to be a Γ-Brownian motion on $\mathbf{M}$.

Approximation theorems for stochastic differential equations justify the assertion that a Γ-Brownian motion can be viewed as the limiting form of a randomly broken geodesic.

DEFINITION (Γ-martingales). If Y is a martingale on $\mathbf{V}$ then X is said to be a Γ-martingale on $\mathbf{M}$.

The importance of Γ-martingales is exemplified in the section on harmonic maps. Note that a Γ-Brownian motion is a special case of a Γ-martingale. A somewhat degenerate example is the 'Brownian-time Γ-geodesic', $\gamma(B)$, where γ is a Γ-geodesic and B is a BM(R).

DEFINITION (Γ-martingales of bounded dilatation). If Y is a martingale on $\mathbf{V}$ of bounded dilatation, then X is said to be a Γ-martingale of bounded dilatation.

A Euclidean martingale Y is of bounded dilatation if a naturally associated random process Q of quadratic forms yields forms which are of bounded dilatation. The form Q is defined by

$$d[Y^i, Y^j] = Q^{ij} \Sigma_i^m d[Y^r, Y^r] \qquad (3.5)$$

where $Y_1, \ldots, Y_n$ are the coordinates of Y. If it is possible to choose this Q so that its first two eigenvalues are of uniformly bounded ratio then Q and Y, and hence X, are all of bounded dilatation. Thus, this condition ensures that there are at least two substantial 'directions of randomness' in the Γ-martingale. It therefore forces the Γ-martingale to behave more like a Brownian motion than the degenerate 'Brownian-time Γ-geodesic' example mentioned above.

Of course, these four examples do not exhaust all the possibilities. Bounded dilatation can be replaced by a stronger condition such as bounded quasi-conformality. The connection may be chosen to reflect some symmetry, or a complex structure (see, for example, the discussion and references to Kahler diffusions in Ikeda and Watanabe, 1984). In each case new possibilities arise. The general program of stochastic differential geometry is to explore the properties and geometrical implications of all of these families of random processes on manifolds.

3.3. CALCULATIONS WITH Γ-BROWNIAN MOTION

We now turn to the special case of Γ-Brownian motion X driven by Euclidean Brownian motion B. Thus, (3.3) takes the form

$$d_S X = \Xi d_S B, \qquad d_S \Xi = H_\Xi \, d_S X \tag{3.6}$$

where B is BM(**V**). Calculations concerning X generally begin with an Itô analysis of the semimartingale $f(X)$ for some suitable smooth real-valued function f. This is closely related to the martingale characterization (Stroock and Varadhan, 1979).

It is actually more efficient first to analyze $g(\Xi)$ for g a smooth real-valued function on $O(\mathbf{M})$, and then to specialise by setting $g = f \circ \pi$. Calculation of Stratonovich differentials yields

$$\begin{aligned} d_S g(\Xi) &= (Tg) \, d_S \Xi \\ &= (Tg) H_\Xi \Xi \, d_S B \end{aligned}$$

using (3.6). Choosing an orthonormal basis $\{w_i : i = 1, \ldots, m\}$ for **V** we set

$$B = \Sigma_i B^i w_i \tag{3.7}$$

and deduce

$$d_S g(\Xi) = \Sigma_i L(w_i) g(\Xi) \, d_S B^i \tag{3.8}$$

where

$$L(w_i)(\xi) = H_\xi \xi w_i \tag{3.9}$$

is the horizontal tangent vector field on the orthonormal frame bundle corresponding to the vector w_i. The relationship between Γ-Brownian motion and Γ-geodesics is indicated by

$$L(w_i) g(\xi) = \frac{d}{dt} \, g(\xi_i(t)) \,|_{t=0} \tag{3.10}$$

where ξ_i is the horizontal lift of the Γ-geodesic with initial velocity $\xi \cdot w_i$.

To obtain the Itô differential corresponding to (3.3), we must find the bracket processes linking B to $L(w_i)g$. From the remark at the end of Section 1, we need only find the Stratonovich differential of $L(w_i)g$. But this last is simply a smooth real-valued function on the orthonormal frame bundle and so the analysis above carries over directly. Consequently

THEOREM 4 (Itô analysis of Γ-Brownian motion).

$$d_I g(\Xi) = \Sigma_i L(w_i) g(\Xi) \, d_I B^i + \tfrac{1}{2} \Sigma_i L^2(w_i) g(\Xi) \, dt \tag{3.11}$$

where

$$L^2(w_i)g(\xi) = \frac{d^2}{dt^2} g(\xi_i(t))|_{t=0} \tag{3.12}$$

where ξ_i is as above.

REMARK. An entirely similar analysis carries through for Γ-martingales. Mixed second derivatives may then appear in the locally-bounded-variation term in which, of course, the dt must be replaced by the bracket processes of the stochastic development martingale. The analysis even extends to general semi-martingales, but then the first term is no longer purely a local martingale differential.

An equivalent form of the Itô analysis theorem states that for all smooth real-valued functions g of compact support on $O(\mathbf{M})$ the Itô differential

$$d_I g(\Xi) - \tfrac{1}{2}\Sigma_i L^2(w_i)g(\Xi)\,dt \tag{3.13}$$

is that of a martingale. Thus the parallel transport Ξ is associated to the horizontal Laplacian on $O(\mathbf{M})$, namely

$$\tfrac{1}{2}\Sigma_i L^2(w_i)$$

while X is associated with the Γ-Laplacian on $\mathbf{M}$ (with respect to the prescribed connection Γ) given by

$$\tfrac{1}{2}\Delta f(x) = \tfrac{1}{2}\Sigma_i L^2(w_i)(f \circ \pi)(\xi) \tag{3.14}$$

when $\pi\xi = x$. Because of the equivariance condition on H, the Γ-Laplacian Δ is well-defined.

The relationship between Γ-Brownian motion and Γ-geodesics is encapsulated in the useful formula for the Γ-Laplacian

$$\tfrac{1}{2}\Delta f(x) = \tfrac{1}{2}\Sigma_i \, d^2/dt^2 f(\gamma_i(t))|_{t=0} \tag{3.15}$$

where the γ_i form an orthonormal set of Γ-geodesics emanating from x.

3.4. THE LEVI–CIVITA CONNECTION AND BM(M)

At the time of writing, most work in stochastic differential geometry considers Γ-Brownian motions and Γ-martingales using the special connection associated uniquely to a Riemannian manifold $\mathbf{M}$, the Levi–Civita connection. In technical terms this is the unique connection on $O(\mathbf{M})$ which is torsion-free: the *torsion form* is identically zero. We shall not define torsion here but remark only that zero torsion is characterized by

$$[L(u), L(v)]f \circ \pi = (L(u)L(v) - L(v)L(u))f \circ \pi = 0 \tag{3.16}$$

for all smooth f on $\mathbf{M}$ and all u and v in $\mathbf{V}$. In this torsion-free case we refer to *the* Brownian motion BM(**M**) on **M**, and to the Γ-Laplacian as the *Laplace–Beltrami operator* Δ on **M**. When necessary to distinguish between Laplace–Beltrami operators on different manifolds, we will use a superfix thus

$^M \Delta$ is the Laplace-Beltrami operator on **M**.

There is a rich general theory of connections. For the purposes of research in stochastic differential geometry, it is necessary to investigate the theory in more depth than can be afforded here, and especially to be familiar with the relationship between connections as presented here and covariant differentiation. The reader is referred to the reference work Kobayashi and Nomizu (1963), to Cheeger and Ebin (1975), and to the excellent introductory text of Bishop and Crittenden (1964).

The remainder of this paper deals entirely with the processes arising from a Levi–Civita connection. The more general treatment above is useful for three reasons:

(1) the class of elliptic diffusions with smooth coefficients can be represented by Γ-Brownian motions using general connections (Ikeda and Watanabe, 1981 page 271, prop. 4.3, and Darling, 1982);
(2) the general treatment clarifies the fundamental notion of stochastic parallel transport;
(3) moreover it is suggestive of generalizations to cases in which there is no Riemannian structure, or the metric structure is not of first importance.

3.5. OTHER CONSTRUCTIONS OF BM(**M**)

There are many different ways of constructing BM(**M**). That they lead to the same random process can be demonstrated in each case by verifying the basic martingale characterization (3.8). Historically, the first general construction was based on operator semigroup methods; BM(**M**) being defined as the Markov process with infinitesimal generator the Laplace–Beltrami operator (together with specification of the correct domain of the operator). The martingale characterization falls very naturally into this circle of ideas. Further developments discuss BM(**M**) as the limiting case of various random walks; see, for example, Roberts and Ursell (1960) and Pinsky (1983).

A more explicit but less intrinsic construction patches together solutions to locally-defined Itô stochastic differential equations (Itô, 1950, McKean, 1969). The construction we have given avoids patching by working on the frame-bundle. A related construction uses 'product-integral injection' (Gangolli, 1964; McKean, 1969; Darling, 1984b).

If the manifold is isometrically embedded in Euclidean space, then BM(**M**) may be constructed as a solution to a stochastic differential equation using Brownian motion on the ambient Euclidean space; the stochastic differential equation depends, of course, on the particular embedding. Nash's theorem asserts that such an isometric embedding is always possible so long as the ambient dimension is sufficiently large. Van Den Berg and Lewis (1985) discuss an interesting special case.

The formalism of second-order differential geometry allows a direct interpretation of the stochastic differentials $d_S X$ and $d_I X$; this is discussed, for example, in Schwartz (1984) and Meyer (1981).

4. Geodesics and Curvature

We have defined geodesics to be solutions of the differential system

$$d_S \gamma_v = \xi(v\, dt), \qquad d_S \xi = H_\xi\, d_S \gamma_v \tag{4.1}$$

here $\gamma'_v(0) = v$ and $\gamma_v(0) = x$. A Riemannian manifold is said to be complete if for the Levi–Civita connection the geodesic equations have solutions for all time for all initial conditions. In this case the map

$$\mathrm{Exp}_x : v \mapsto \gamma_v(1)$$

defines the exponential map at x, mapping the tangent fiber at x onto the whole of the manifold. The exponential map at x is smooth and is invertible when $\|v\|$ is suitably small. Hence, it provides an important system of coordinates around x, the normal coordinate system

$$(r, \theta) = (\|v\|, v/\|v\|) \mapsto \gamma_v(1)$$

In the system of normal coordinates we may measure lengths of curves by the classical formula

$$ds^2 = dr^2 + G(r, \theta)^2\, d\theta^2 \tag{4.2}$$

where $G(r, \theta)$ is a positive-definite symmetric matrix. In the particular case of the sphere of radius $1/k$, this matrix $G(r, \theta)$ is diagonal, with diagonal elements all equal to $k^{-1} \sin kr$. For small r, the point at (r, θ) is at distance r from the base point x, but for large r this may not be so any more. The reason is that there can be an alternative coordinate (r', θ') with $r' < r$ and

$$\mathrm{Exp}_x(r', \theta') = \mathrm{Exp}(r, \theta)$$

This phenomenon occurs on the sphere (obviously, since no two points are separated by more than π/k if the sphere is of radius $1/k$). We set

$$f_c(\theta) = \inf\{r : \mathrm{Exp}_x(r, \theta) \text{ and } x \text{ are closer than } r\} \tag{4.3}$$

and define the cut-locus at x as

$$C(x) = \{\mathrm{Exp}_x(r, \theta) : r = f_c(\theta)\} \tag{4.4}$$

Thus, the cut-locus $C(x)$ marks out the maximal region in which a normal coordinate system can be employed. The region

$$U(x) = \{(r, \theta) : r < f_c(\theta)\} \tag{4.5}$$

is a star-shaped open region in $T_x\mathbf{M}$ and is mapped smoothly by Exp_x onto $\mathbf{M}\text{-}C(x)$ (which itself is an open dense subset of $\mathbf{M}$). Since $T_x\mathbf{M}$ is a Euclidean space, our original intuitive picture of a manifold, as an open subset of a Euclidean space together with identifications at the boundary, is entirely correct! However, the identifications can be very complicated; moreover the $U(X)$ for different x will not, in general, mesh well together.

Fix p in $\mathbf{M}$ once for all as base point. The distance function

$$r(x) = \text{distance}(x, p) \tag{4.6}$$

is fundamental in stochastic differential geometry. Many results about BM($\mathbf{M}$) depend on an analysis of $r(X)$ where X is BM($\mathbf{M}$). The function r is smooth on

$$\mathbf{M} - C(p) - \{p\}$$

and X almost surely never hits p at positive times if dim $\mathbf{M} > 1$, so the stochastic calculus can be applied to calculate $d_I r(X)$ until X first hits $C(p)$. The resulting analysis uses Theorem 4 and yields

$$d_I r(X) = \text{grad } r(X) \Xi \, d_I B + \tfrac{1}{2} \Delta r(X) \, dt \tag{4.7}$$

(valid only till the time that X first hits $C(p)$). The gradient grad r and the Laplacian Δr satisfy the following equations, which indicate how Theorem 4 can be used in their proofs:

$$\text{grad } r(x) = (L(w_1)r \circ \pi, \ldots, L(w_n)r \circ \pi)(\xi) \tag{4.8}$$

(where $y \cdot \xi = x$, so that ξ is a vector in $T_x\mathbf{M}$, and $w_1, \ldots, w_n$ form an orthonormal basis for $\mathbf{R}^m$);

$$\Delta r(x) = \Sigma \, L^2(w_1)(r \circ \pi)(\xi) \tag{4.9}$$

Information about grad r and Δr is available by arguments from nonstochastic differential geometry, as for example in Cheeger and Ebin (1975). Gauss' lemma implies that grad r is of length 1 and so by Levy's characterization (1.11) the process

$$W_t = W_0 + \int \text{grad } r(X) \Xi \, d_I B \tag{4.10}$$

is a one-dimensional Brownian motion stopped when X first hits $C(p)$ (here B is the stochastic development and Ξ the parallel transport of X). The analysis of Δr is deeper and depends on the notion of curvature. This is the crux of stochastic differential geometry; all we can give here is a brief intuitive discussion and a reference to Cheeger and Ebin (1975) or Milnor (1963) for rigorous treatment.

4.1. CURVATURE

Consider two particles moving at constant speed along geodesics originating from a common point but issuing at slightly different angles. To the first order the distance between the particles increases at a linear rate for small times. Curvature measures the second-order correction to this: the sign of the curvature is the opposite of the sign of the second-order correction term. Hence, a sphere has positive curvature (and in that case the particles will meet at the point antipodal to the starting point). It is clear from this description that the curvature function essentially depends on the originating point and on two directions at that point. Some remarkable theorems in differential geometry elicit strong global conclusions from hypotheses on the curvature function. For example:

THEOREM 5 (Cartan–Hadamard). *If **M** is simply connected and all its curvatures are nonpositive, then the cut-loci are empty and, hence, the exponential maps provide global coordinate systems. In particular, as a smooth manifold **M** is like Euclidean space (albeit with a different metric).*

If ξ is a frame above the originating point and the two geodesics in question have tangents $L(u)(\xi)$ and $L(v)(\xi)$ then the curvature is quantified by the component of

$$-[L(u), L(v)] = -L(u)L(v) + L(v)L(u)$$

that is perpendicular to the horizontal vectors of $T_\xi O(\mathbf{M})$. From this is derived the curvature transform $R(u, v)w$, which is defined for vectors u, v, w in $T_x\mathbf{M}$ and yields a vector in $T_x\mathbf{M}$. This in turn is used to define two important entities:

(1) the sectional curvature $K(P)$ of the tangent plane P spanned by the orthonormal pair of tangent vectors u and v:

$$K(P) = \langle R(u, v), u, v \rangle$$

(2) the Ricci curvature quadratic form ρ of the direction u:

$$\rho(u) = \Sigma \langle R(u, w_i)u, w_i \rangle$$

where the w_i form an orthonormal basis of $T_x\mathbf{M}$.

In the particular case when **M** is rotationally symmetric about p, Equation (4.2) for the metric becomes

$$ds^2 = dr^2 + g(r)^2 \, d\theta^2 \tag{4.11}$$

where g is a scalar function depending only on r with $g(0) = 0$ and $g'(0) = 1$. Then **M** is a model in the sense of Greene and Wu (1979). The symmetry reduces many problems to questions about one-dimensional ordinary differential equations. In particular, the sectional curvature $K(r)$ in any direction containing the tangent vector grad r must satisfy the one-dimensional Jacobi equation

$$g''(r) = -K(r)g(r). \tag{4.12}$$

Thus, any prescribed $K(r)$ can be attained by suitable choice of g (this stands in strong contrast to the general case, although even in the case of a model, the nonradial sectional curvatures are in general uncontrolled). Moreover, the Laplace–Beltrami operator for a model satisfies

$$\Delta r = (m-1)g'(r)/g(r) \tag{4.13}$$

It is possible to compare Laplace–Beltrami operators as applied to the distance functions arising from a general manifold and from a model: this allows comparisons between the stochastic differential geometry of a general manifold and the stochastic differential geometry of a model.

THEOREM 6 (comparison theorem for Laplace–Beltrami operators). *Suppose* **M** *is a model with rotational symmetry about p and* **N** *is a manifold of the same dimension such that for some q in* **N**

Ricci (*x, radial direction*)

$\leqslant$

Ricci (*y, radial direction*)

whenever x is in **M**-$C(p)$-$\{p\}$ *and y is in* **N**-$C(q)$-$\{q\}$ *and*

$$r_M(x) = r_N(y).$$

Here r_M *and* r_N *are the distances in the respective manifolds from p and q, respectively. Then*

$$^M\Delta r_M(x) \geqslant {}^N\Delta r_N(y) \tag{4.14}$$

This result can be seen to be plausible on considering the geodesic interpretations of curvature above and of the Laplace–Beltrami operator; see (3.10). Moreover, the geodesic interpretation of the Laplace–Beltrami operator allows one to reduce the calculation of $^M\Delta r_M$ to an essentially one-dimensional problem.

A similar theorem deals with the more general case when neither **N** nor **M** is a model. The curvature condition must then be strengthened to

each sectional curvature at x for a radial plane

$\leqslant$

all sectional curvatures at y for radial planes.

Moreover, the comparison can then be made for Hessians rather than Laplace–Beltrami operators. This is an important point in applications to the theory of harmonic maps, in which it becomes necessary to study $r(X)$ when X is a Γ-martingale of bounded dilatation.

5. Properties of BM(M)

With the machinery that has been set up, relationships can be established between the behaviour of BM(**M**) and the geometrical properties of **M**. The method of attack is an Itô analysis of the real-valued random process which is the radial part of BM(**M**)

$$r(X) = \text{distance}(X, p)$$

followed by application of the geometrical relationships between curvature and the Laplace–Beltrami operator indicated in the previous section, and then of comparison theorems for one-dimensional stochastic differential equations, such as Theorem 3 above. The method generalizes readily to the case when X is a Γ-martingale; this is important in the theory of harmonic maps as discussed in the following section.

5.1. EXPLOSIONS AND STOCHASTIC COMPLETENESS

If the Ricci curvatures of **M** do not become too large and negative then BM(**M**) will not explode. So in such cases **M** is stochastically complete. For our purposes the most attractive way to see this is to use Itô calculus to compare the Brownian path with the path of a Brownian motion on a special manifold.

THEOREM 7 (Ducourtioux, 1983, and Ichihara, to appear; see also Azencott, 1974, and Yau, 1978). *If the Ricci curvatures of* **M** *are bounded below by a negative quadratic function of the distance function* r, *then* **M** *is stochastically complete.*

   Proof. The quadratic lower bound implies that for some constant C

$$\text{Ricci}(x) \geqslant -C(3 + Cr^2). \tag{5.1}$$

Suppose **N** is a model of the same dimension as **M** and with radial sectional curvatures prescribed by the lower bound above. The metric of **N** is given by

$$ds^2 = dr^2 + r^2 \exp(Cr^2) \, d\theta^2 \tag{5.2}$$

and its Laplace–Beltrami operator can be shown to satisfy

$${}^N\Delta r = (m-1)(r^{-1} + Cr) \tag{5.3}$$

Theorem 6 shows

$${}^M\Delta r \leqslant {}^N\Delta r \text{ off the cut-locus } C(p). \tag{5.4}$$

If there is no chance of X hitting the cut-locus $C(p)$ (if the cut-locus is polar) then Equation (4.7) can be used together with the comparison Theorem 3 to show that $r(X)$ is bounded above by Y where Y solves

$$d_I Y = d_I W + (m-1)(1/(2Y) + CY/2) \, dt. \tag{5.5}$$

But Y can in turn be bounded above by Z, which is obtained from the 'unstable Ornstein–Uhlenbeck process'

$$e^{kt} r(X_0) + k \int \exp(k(t-s)) \, W_s \, ds + W_t \tag{5.6}$$

(which solves $d_I Z = d_I W + kY \, dt$) so long as we neglect the $(m-1)/(2Y)$ term in the drift of Y; here

$$k = (m-1)(4/r(X_0)^2 + C)/2.$$

The process at (5.6) is clearly nonexplosive; a further comparison argument shows that despite the neglected term, the process Y does not explode either. Comparison then shows that $r(X)$ cannot explode.

If X might hit the cut-locus then application must be made either of a smoothing argument or of the results of Kendall (preprint), which show how to extend (4.7) for all time.

REMARK. If the cut-locus is polar and

$$\text{Ricci}(x) \leqslant -r^{2+\epsilon} \text{ for some positive } \epsilon \qquad (5.7)$$

then a similar approach shows that BM(**M**) will explode. Tighter results are available; see, for example, Li and Schoen (1984).

5.2. TRANSIENCE AND RECURRENCE

For X to be recurrent it must return to a fixed open set at arbitrarily large times. This is equivalent to

$$\liminf_{t \to \infty} r(X) = 0 \qquad (5.8)$$

since if the limit infimum is nonzero, then Markov process arguments show that it must be infinite; hence, X must be transient and travel off to infinity. Recurrence is equivalent to the condition that all bounded subharmonic functions are constant (if BM(**M**) is transient then the quantity which is the expected future time spent in a specified set will provide such a function).

Simple comparison arguments can be applied as above to show that in dimension 2 all nonnegatively curved manifolds have recurrent Brownian motion, whereas in higher dimensions all nonpositively curved manifolds have transient Brownian motions. For a properly detailed picture, a more subtle analysis is necessary. Theorem 4' of Lyons and Sullivan (1984) is a criterion due to Kelvin, Nevanlinna, and Royden giving a necessary and sufficient condition for transience in terms of the existence of a vector field of nonzero but finite divergence and finite energy. Durrett (1986) surveys results due to many workers (among whom Ichihara and Varopoulos give approachs oriented towards probability theory) which connect transience with conditions on the growth of geodesic balls in the manifold.

5.3. ZERO-ONE LAWS

If X is transient, then its asymptotic behaviour is of interest. This is tied to the existence of nonconstant bounded harmonic functions on **M**. For suppose that $\mathcal{P}$

is an asymptotic property of X. Then the function

$$f(x) = P\{X \text{ has property } \mathcal{P} \mid X_0 = x\}$$

is a bounded harmonic function.

THEOREM 8 (Zero-One Law for nonnegative Ricci curvature). *If the Ricci curvatures of* **M** *are nonnegative, then* X *satisfies a* 0–1 *law; which is to say all bounded harmonic f are constant and asymptotic properties of* X *therefore have probability* 0 *or* 1.

This theorem was proved probabilistically by Debiard, Gaveau, and Mazet (1976) subject to the assumption that **M** is stochastically complete (but we have seen above that such **M** are always stochastically complete). A more general theorem can be found with an analytic proof in Yau (1975), and further generalizations allowing weak negative curvature are given in Greene and Wu (1979).

If **M** is compact or, more generally, if its Brownian motion is recurrent, then it is immediate that X satisfies a 0–1 law. However, the same is not necessarily true for a covering manifold of **M**; see for example the work of Lyons and Sullivan (1984).

5.4. LIMITING DIRECTIONS

If X does not satisfy a 0–1 law, then its asymptotic behaviour has a random component. If X is of constant negative curvature and simply connected, then its asymptotic behaviour includes a limiting direction randomly distributed over the absolute sphere of directions; Dynkin (1961), Orihara (1970) and Norris *et al.* (1986) have covered this and also the question of more general noncompact symmetric spaces (direction here is interpreted in the sense of polar coordinates as in Theorem 5). In the case of constant negative curvature a simple stochastic proof can be constructed following the method used in the two-dimensional case by Kendall (1984); one studies the distance of X from a totally geodesic hypersurface.

Considerable work exists on the case when **M** has variable negative curvature bounded away from zero; see Kifer (1976), Prat (1975), Kendall (1984), Darling (1985), and Hsu and March (1985). Kifer (1976), Sullivan (1983), Anderson (1983), and Anderson and Schoen (1985) discuss implications for the representation of bounded (or, more generally, positive) harmonic functions. We content ourselves here by informally discussing Prat's result, which has proved seminal in the probabilistic approach to harmonic map theory.

THEOREM 9 (Prat, 1975). *Let* **M** *be a simply-connected manifold with 'pinched' sectional curvatures*

$$-L^2 \leqslant \text{Sect}(\mathbf{M}) \leqslant -K^2 < 0.$$

If Θ is the angular part of BM(**M**),

$$X = \mathrm{Exp}_p(R, \Theta)$$

then Θ converges to a limiting value.

Proof. The lower bound on curvature ensures that X does not explode. Comparison arguments show that as t tends to ∞ so

$$\liminf r(X_t)/t \geqslant (m-1)K/2,$$
$$\limsup r(X_t)/t \leqslant (m-1)L/2.$$

Thus, X diverges to infinity. But the negative curvature means that the surface measure of the geodesic ball $\{x : r(x) = \rho\}$ increases exponentially fast with ρ and the effect of this is that the diffusive component of Θ decreases rapidly. The drift of Θ can be controlled by techniques related to the existence of the lim sup bound and so Θ can be shown to 'freeze to a halt'.

It is straightforward to verify that the limit is truly random (see Sullivan, 1983; also the harmonic map applications in Kendall, 1981, 1983; Goldberg and Mueller, 1983). Most of the work since Prat has been concerned with weakening the upper bound on the curvature and, above all, with lifting the annoying lower bound at least in special cases. It still appears to be an open problem whether some kind of lower bound on the curvature is necessary; however see Ancona (1985) for a new approach.

The effect of a limiting direction is to localize the influence of the metric on the behaviour of BM(**M**). Pinsky (1977) exploits this to prove an individual ergodic theorem for each Brownian path on manifolds such as **M** in Theorem 9.

6. Brownian Motion and Harmonic Maps

There is a remarkable relationship (first observed by P. Levy) between BM($\mathbf{R}^2$), viewed as complex Brownian motion, and complex analysis; complex Brownian motion under a holomorphic map is essentially a new complex Brownian motion (but subject to a random change of time). Together with a similar result relating BM(**M**) to harmonic functions (traces of which can be noted in the discussion of the previous section), this has lead to a fruitful collaboration between analysis and probability. See, for example, the encyclopaedic text of Doob (1984) and also Durrett (1984).

The relationship generalizes naturally to maps between manifolds **M** and **N**.

THEOREM 10 (Fuglede, 1978). *The map $F : \mathbf{M} \to \mathbf{N}$ sends* BM(**M**) *into a time-changed* BM(**N**) *if and only if these two conditions hold*:

(a) *F is a harmonic map,*

(b) *F is horizontally semi-conformal.*

We then say F is a harmonic morphism.

We will define the notion of a harmonic map below. For geometric reasons condition (b) is restrictive except in the case of dim $\mathbf{N} = 2$. In other cases, harmonic morphisms are rigid and relatively rare entities, while maps that merely satisfy condition (a) are more common. The surveys of Eells and Lemaire (1978, 1983) are valuable sources of information about both kinds of map.

6.1. HARMONIC MAPS AND Γ-MARTINGALES

A *harmonic map* $F: \mathbf{M} \to \mathbf{N}$ between Riemannian manifolds $\mathbf{M}$ and $\mathbf{N}$ can be characterized by a variational formulation, or a nonlinear differential equation, or potential theory, or a semimartingale property. For us, the probabilistic definition is most convenient; F is harmonic if and only if $F(\mathrm{BM}(\mathbf{M}))$ is a Γ-martingale on $\mathbf{N}$. This characterization is due to Bismut; see Meyer (1981) and also Darling (1982).

All the characterizations are of course equivalent, though different approaches suggest different problems and generalizations. To date, existence results have not been obtained by the probabilistic approaches; the probabilistic construction of harmonic maps is an open problem (but see Ducourtioux, 1976, 1978). However, semimartingale methods are fruitful in providing nonexistence results. The following example sets the scene.

THEOREM 11. *Suppose that* $\mathbf{N}$ *is simply-connected with* $Sect(\mathbf{N}) \leq 0$. *If all bounded harmonic functions on* $\mathbf{M}$ *are constant, then all harmonic maps* $F: \mathbf{M} \to \mathbf{N}$ *of bounded image are also constant.*

Proof. The most direct proof is potential-theoretic. It can be expressed in probabilistic terms. Let $Y = F(\mathrm{BM}(\mathbf{M}))$. An Itô analysis of the distance of Y from a fixed point shows that the distance process is a submartingale. By hypothesis it is bounded, and so must converge. Such convergence (for arbitrary fixed points) implies that Y itself must converge to a limit $Y(\infty)$ which must be nonrandom (or else we can construct nonconstant bounded harmonic functions). But then the distance of Y from this point is a bounded submartingale that converges to zero. Basic martingale theory asserts that all such submartingales are constant.

Darling (1985) uses the general theory of Γ-martingales to produce a similar proof.

The above theorem illustrates the basic strategy as propounded in Kendall (1981): one can deduce geometric implications in harmonic map theory by contrasting properties of $\mathrm{BM}(\mathbf{M})$ with properties of families of Γ-martingales in $\mathbf{N}$. Equally, a property of Brownian motion or of Γ-martingales that leads to such implications is thereby interesting and worthy of further study.

6.2. LIMITING DIRECTIONS

Theorem 9 can be mimicked for Γ-martingales of bounded dilatation. The next theorem follows by a lifting argument and by arguing as in Theorem 11 above.

THEOREM 12 (Kendall, 1981, 1983). *Suppose that* **N** *satisfies*

$$-L^2 < Sect(\mathbf{N}) < -K^2 < 0$$

and is simply connected. If all bounded harmonic functions on **M** *are constant then all harmonic maps* $F: \mathbf{M} \to \mathbf{N}$ *of bounded dilatation are also constant.*

REMARK. As described in Kendall (1981), the basic idea of this proof goes back to Davis' (1975) probabilistic proof of Picard's little theorem of complex analysis.

Theorem 12 was originally proved by geometric methods, employing the stronger condition Ricci(**M**) ≥ 0, but dispensing with the lower curvature bound on **N**. The geometric approach actually allows a trade-off between negative Ricci curvatures on **M** and the upper curvature bound on **N**. On the other hand, Goldberg and Mueller (1983) show that the probabilistic proof allows for small regions of positive curvature and (not too rapid) decay of negative curvature to zero at infinity on **N**, so long as the bounded dilatation condition is replaced by bounded quasi-conformality. So the probabilistic approach deals with the asymptotic behaviour of **M** and **N** and, thus, differs fundamentally from the geometric approach.

6.3. THE BROWNIAN COUPLING PROPERTY

The irritating lower curvature bound of Theorem 9 appears again above. We must take a completely different approach in order to dispense with this.

DEFINITION. The manifold **M** has the *Brownian coupling property* (BCP) if for any two distinct points x, y in **M** we may construct X, Y to be semimartingales begun at x, y, respectively, which are Brownian motions with respect to the same filtration, and which meet almost surely, or *couple*.

REMARK. A weaker coupling property not requiring the filtration to be the same is exploited by Lyons and Sullivan (1984). Their property is directly linked to the constancy of bounded harmonic functions. The potential-theoretic implications of the BCP are currently under investigation; the BCP clearly implies constancy of all bounded anti-spacetime-harmonic functions.

CONJECTURE. If all bounded anti-spacetime-harmonic functions are constant, then **M** has the BCP.

The work of Lindvall and Rogers (1986) shows that Euclidean spaces possess the BCP by using a 'half-of-mirrors' construction. Kendall (1986b) establishes the BCP for Riemannian manifolds with nonnegative Ricci curvatures (and also for all compact manifolds).

Following on the fundamental Lindvall–Rogers result the BCP is of significance in stochastic differential geometry because of the following.

THEOREM 13 (Kendall, 1986a). *Suppose that* **N** *satisfies*

$$Sect(\mathbf{N}) < -K^2 < 0$$

and is simply connected. If X, Y *are* Γ-*martingales of bounded dilatation on* **N** *with* X_0 *and* Y_0 *not equal, then the probability that they never meet is positive.*

Proof. This follows the now-familiar lines of Itô analysis of a distance, namely

distance(X, Y).

A comparison argument shows it is greater than a Brownian motion of constant positive drift. The result follows by (1.8).

One geometric implication is immediate.

THEOREM 14. *If* **M** *has the BCP and* **N** *is as in the previous theorem, then every harmonic map of bounded dilatation*

$$F: \mathbf{M} \to \mathbf{N}$$

is constant.

Proof. Otherwise the BCP could be used to provide Γ-martingales beginning at different points in **N** that nevertheless coupled.

The BCP is usefully flexible for geometric applications.

THEOREM 15. *If* **F** *has the BCP,* **N** *is as in the previous theorem, and* **M** *is a general manifold, then every harmonic map of bounded dilatation* $F: \mathbf{M} \times \mathbf{F} \to \mathbf{N}$ *factorizes as* $F = G \circ \pi$ *where* $G: \mathbf{M} \to \mathbf{N}$ *is harmonic and* $\pi: \mathbf{M} \times \mathbf{F} \to \mathbf{M}$ *is the projection map.*

Proof. The BCP can be used to provide Brownian motions on the product that move above the same process on **M** and which couple. By arguing as above it is seen that F is constant on the fibres of π, and the rest follows.

REMARK. The product $\mathbf{M} \times \mathbf{F}$ has the product Riemannian structure.

Elworthy and Kendall (1986) generalize this result by replacing $\mathbf{M} \times \mathbf{F}$ with **E**. Here **E** is a fibre space:

$$\pi: \mathbf{E} \to \mathbf{M}$$

is required to be a Riemannian submersion of totally geodesic fibres. Moreover, the isometric fibres **F** must have the BCP. In the course of this work, a remarkable factorization is found for BM(**E**) in terms of Brownian motion on a typical fibre and a process of fibre isometries built on top of a BM(**M**). This factorization is related to the results of Price and Williams (1983) about BM(S^2).

7. Other Topics

This paper is not intended as an exhaustive survey of stochastic differential geometry but more in the way of an introduction. However, we conclude by giving at least some indication of the variety of work going on in the subject, interpreting the term 'stochastic differential geometry' in a fairly broad sense. In this last section brief summaries are given on each of a number of topics, together with a few references. The references certainly do not form an exhaustive list, and apologies are offered to those who are neglected here.

7.1. LOCAL STOCHASTIC RIEMANNIAN GEOMETRY

An important aspect of Riemannian geometry is the study of geometric properties that rigidly determine Riemannian manifolds. Pinsky (1983), Gray and Pinsky (1983) discuss characterizations of Riemannian manifolds in terms of asymptotics of the mean exit time from a small geodesic ball.

7.2. Γ-MARTINGALES

In this paper, Stratonovich differentials have continually been treated in a formal sense as tangent vectors. Schwartz's approach to stochastic differential geometry uses second-order differential geometry (see Schwartz, 1984, and Meyer, 1981) to carry this beyond a formal correspondence. This approach studies Γ-martingales as objects of their own intrinsic importance. See Darling (1982, 1983, 1984a, 1985), Zheng (1983), and Emery and Zheng (1984).

7.3. STOCHASTIC DIFFERENTIAL FORMS

Stochastic differential 1-forms may be integrated over the path of a Brownian motion. The work of Malliavin (1974) has been seminal here as throughout stochastic differential geometry. Bismut (1981) and Manabe (1982) carry the study further in various directions. Pitman and Yor (1984) take up the work of Lyons and McKean (1984) on the winding of complex Brownian motion. Darling (1984a) uses stochastic differential forms to study geodesic deviation.

7.4. STOCHASTIC FLOWS

An important aspect of stochastic differential systems of manifolds is the dependence of the solution on the initial data. We have referred briefly to this in Section 1. Detailed investigations have been undertaken by Bismut (1981), Elworthy (1978), Kifer (1984), Kunita (1984), Carverhill and Elworthy (1983), and Baxendale (1980). The study leads on naturally to that of the next heading.

7.5. MALLIAVIN CALCULUS

A substantial part of nonstochastic differential geometry can be viewed as the study of variations of geodesics through one-parameter families. In the same way one can consider the effect of varying the Brownian path; this is the Malliavin calculus. Complicated technicalities arise even in the Euclidean case but, nevertheless, the rewards are great. Here we refer only to a recent survey by Ikeda and Watanabe (1983).

7.6. SMALL-TIME ASYMPTOTICS

Varadhan (1969) pointed out that large deviations of Brownian motion over a short time look like geodesics. Further investigations have been extensive; this is very much a current topic of research and involves difficult analysis. See Molchanov (1975), Elworthy and Truman (1982), Watling (in preparation), Bismut (1984a), Azencott *et al* (1981). Something of the depth of this can be gauged by the application made by Bismut (1984b) of the ideas of small-time asymptotics and Malliavin calculus to the Atiyah-Singer index theorem.

7.7. RANDOM WALKS

BM(**M**) is closely related to certain random walks on **M** which may be thought of as discrete approximations to Brownian motion. Their properties are closely tied to those of BM(**M**). See, for example, Lyons and Sullivan (1984) and Varopoulos (1984).

7.8. WIENER SAUSAGES

The Wiener sausage is the tube-like set formed as the union of small geodesic balls of constant radius centred on each point of the trajectory of a Brownian motion. Spitzer (1964) discovered a relationship with electrostatic capacity in the case of Euclidean space; Chavel and Feldman (1986) generalize this to Riemannian manifolds.

7.9. STOCHASTIC KAHLERIAN GEOMETRY

We have mentioned above the close relationship between complex Brownian motion and complex analysis. This carries over to the case of a Kahler manifold (a manifold with Riemannian metric closely tied to a complex structure). See Ikeda and Watanabe (1984, section 4), and the references therein to papers by Debiard and Gaveau, and also Fukushima and Okada (1984).

Acknowledgements

Much of the work on this paper was done while visiting the Mathematical Institute of Warwick University, under the auspices of the SERC-funded Year of Stochastic Analysis. My interest in the subject began because of some provocative suggestions by D. G. Kendall; it has been sustained through his encouragement and that of K. D. Elworthy. For this I give them both many thanks.

References

Ancona, A. (1985) Variétés a courbure negative, opérateurs elliptiques, et frontière de Martin, *CRAS* **A301**, 193–196.

Anderson, M. T. (1983) The Dirichlet problem at infinity for manifolds of negative curvature, *J. Diff. Geom.* **18**, 701–721.

Anderson, M. and Schoen, R. (1985) Positive harmonic functions on complete manifolds of negative curvature, *Ann. Math.* **121**, 429–461.

Antonelli, P. L., Chapin, J., and Voorhees, B. H. (1980) The geometry of random genetic drift VI. A random selection diffusion model, *Adv. Appl. Prob.* **12**, 50–58 (see also references therein).

Azencott, R. (1974) Behavior of diffusion semigroups at infinity, *Bull. Sci. Math.* **102**, 193–240.

Azencott, R. *et al.*, (1981) Geodésics et diffusions en temps petit, *Asterisque* 84–85, Soc. Math. de France.

Baxendale, P. (1980) Wiener processes on manifolds of maps, *Proc. Royal Soc. Edinburgh*, **87A**, 127–152.

Bishop, R. and Crittenden, R. (1964) *Geometry of Manifolds*, Academic Press, New York.

Bismut, J.-M. (1981) *Mecanique Aleatoire*, LN Math. **866**, Springer-Verlag, Berlin.

Bismut, J.-M. (1984a) *Large Deviations and the Malliavin calculus*, Birkhauser, Basle.

Bismut, J.-M. (1984b) The Atiyah–Singer theorems: A probabilistic approach I, II, *J. Funct. Analysis* **57**, 56–99 and 329–348.

Carverhill, A. and Elworthy, K. D. (1983) Flows of stochastic dynamical systems – the functional analytic approach, *ZW* **65**, 245–267.

Chavel, I. (1984) *Eigenvalues in Riemannian geometry*, Academic Press, New York.

Chavel, I. and Feldman, E. A. (1986) The Wiener sausage and a theorem of Spitzer in Riemannian Manifolds, in J. Chao and N. Woycynski, (eds.) *Probability and Harmonic Analysis*, Marcel Dekker, New York, pp. 45–60.

Cheeger, J. and Ebin, D. G. (1975) *Comparison Theorems in Riemannian Geometry*, North Holland, Amsterdam.

Chung, K. L. and Williams, R. (1984) *Introduction to Stochastic Integration*, Birkhauser, Basle.

Darling, R. W. R. (1982) Martingales in manifolds – definitions, examples, and behaviour under maps, *in Sem. Prob. XVI (supplement)*, LN Math. **921**, Springer-Verlag, Berlin, pp. 217–236.

Darling, R. W. R. (1983) Convergence of martingales on a Riemannian manifold, *Publ. RIMS Kyoto Univ.* **19**, 753–763.

Darling, R. W. R. (1984a) Approximating Itô integrals of differential forms and geodesic deviation, *ZW* **65**, 563–572.

Darling, R. W. R. (1984b) On the convergence of Gangolli processes to Brownian motion on a manifold, *Stochastics* **12**, 277–302.

Darling, R. W. R. (1985) Convergence of martingales on manifolds of negative curvature, *Ann. Inst. H. Poincaré* **21**, 157–175.

Darling, R. W. R. (to appear) The angular part of Brownian motion as a martingale on the sphere. Preprint.

Davis, B. (1975) Brownian motion and Picard's theorem, *TAMS* **213**, 353–362.

Debiard, A., Gaveau, B., and Mazet, E. (1976) Théorèmes de comparaison en géométrie Riemannienne, *Publ. RIMS Kyoto Univ.* **12**, 391–425.

Dellacherie, C and Meyer, P.-A. (1978) *Probabilities and Potential, A*, North Holland, Amsterdam.

Dellacherie, C and Meyer, P.-A. (1982) *Probabilities and Potential, B*, North-Holland, Amsterdam.

Doob, J. L. (1984) *Potential Theory and its Probabilistic Counterpart*, Springer-Verlag, Berlin.

Ducourtioux, J. (1976) Formule de la moyenne pour les applications harmoniques, *Bull. Sci. Math.* **100**, 229–239.

Ducourtioux, J. (1978) Temps de vie des solutions de l'équation de la chaleur de Eells–Sampson, *CRAS* **A286**, 333–336.

Ducourtioux, J. (1983) Temps de vie du brownien et conditions de courbure, *CRAS* **A296**, 769–772.

Durrett, R. (1984) *Brownian Motion and Martingales in Analysis*, Wadsworth, U.S.A.

Durrett, R. (1986) Reversible diffusion processes, in J. Chao and W. Woyczynski, (eds.), *Probability and Harmonic Analysis*, Marcel Dekker, New York, pp. 67–89.

Dynkin, E. B. (1961) Nonnegative eigenfunctions of the Laplace–Beltrami operator and Brownian motion in certain symmetric spaces, *Dok. Akad. Nauk. SSSR* **141**, 1433–1436.

Eells, J. and Elworthy, K. D. (1970) Wiener integration on certain manifolds, in *Problems in Nonlinear Analysis*, CIME IV, 67–94.

Eells, J. and Lemaire, L. (1978) A report on harmonic maps, *Bull. LMS* **10**, 1–68.

Eells, J. and Lemaire, L. (1983) *Selected Topics in Harmonic Maps*, CBMS regional conference series **50**, AMS, Providence.

Elworthy, K. D. (1978) Stochastic dynamical systems and their flows, in *Stochastic Analysis*, Academic Press, New York, pp. 79–95.

Elworthy, K. D. (1982) *Stochastic Differential Equations on Manifolds*, CUP, London.

Elworthy, K. D. and Kendall, W. S. (1986) Factorization of Brownian motion and harmonic maps, in K. D. Elworthy (ed.), *From Local Times to Global Geometry, Control and Physics*, Pitman Research Notes in Maths, No. 150, pp. 75–83.

Elworthy, K. D. and Truman, A. (1982) The diffusion equation and classical mechanics: an elementary formula, In Albeverio *et al.*, (eds.), *Stochastic Processes in Quantum Physics*, LN Physics **173**, Springer-Verlag, Berlin, pp. 136–146.

Emery, M. and Zheng, W. A. (1984) Fonctions convexes et semimartingales dans une variété, *Sem. Prob. XVIII*, LN Maths **1059**, Springer-Verlag, Berlin, pp. 501–518.

Fuglede, B. (1978) Harmonic morphisms between Riemannian manifolds, *Ann. Inst. Fourier (Grenoble)* **28**, 107–144.

Fukushima, M. and Okada, M. (1984) On conformal martingale diffusions and pluripolar sets, *J. Funct. Anal.* **55**, 377–388.

Gangolli, R. (1964) On the construction of certain diffusions on a differentiable manifold, *ZW* **2**, 406–419.

Goldberg, S. I., Ishihara, T., and Petridis, N. C. (1975) Mappings of bounded dilatation of Riemannian manifolds, *J. Diff. Geom.* **10**, 619–630.

Goldberg, S. I. and Mueller, C. (1983) Brownian motion, geometry, and generalizations of Picard's little theorem, *Ann. Prob.* **11**, 833–846.

Gray, A. and Pinsky, M. A. (1985) The mean exit time from a small geodesic ball in a Riemannian manifold, *Bull. Sci. Math.* **107**, 1–26.

Greene, R. E. and Wu, H. (1979) *Function Theory on Manifolds which Possess a Pole*, LN Math. **699**, Springer-Verlag, Berlin.

Hsu, P. and March, P. (1985) The limiting angle of certain Brownian motions, *Comm. Pure Appl. Maths* **38**, 755–768.

Ichihara, K. (1982) Curvature, geodesics, and the Brownian motion on a Riemannian manifold I, II, *Nagoya Math. J.* **87**, 101–114 and 115–125.

Ichihara, K. (to appear) Comparison theorems for Brownian motions on Riemannian manifolds and their applications, *J. Multivariate Analysis*.

Ikeda, N. and Watanabe, S. (1981) *Stochastic Differential Equations and Diffusion Processes*, North-Holland/Kodansha, Amsterdam and Tokyo.

Ikeda, N. and Watanabe, S. (1983) An introduction to Malliavin's calculus, in *Taniguchi Symposium, Katata* 1983, pp. 1–52.

Ikeda, N. and Watanabe, S. (1984) Stochastic flows of diffeomorphisms, in M. Pinsky, (ed.), *Advances in Probability*, No. 7, Marcel Dekker, New York.

Itô, K. (1950) On stochastic differential equations on a differentiable manifold. 1, *Nagoya Math. J.* **1**, 35–47.

Kendall, D.G. (1977) The diffusion of shape (abstract), *Adv. Appl. Prob.* **9**, 428–430.

Kendall, W. S. (1981) Brownian motion, negative curvature, and harmonic maps, in D. Williams, (ed.) *Stochastic Integrals*, LN Math **851**, Springer-Verlag, Berlin.

Kendall, W. S. (1983) Brownian motion and a generalised little Picard's theorem, *TAMS* **275**, 751–760.

Kendall, W. S. (1984) Brownian motion on a surface of negative curvature, *Sem. Prob. XVIII*, LN Math **1059**, Springer-Verlag, Berlin.

Kendall, W. S. (1986a) Stochastic differential geometry, a coupling property, and harmonic maps, *Proc. LMS*, **33**, 554–566.

Kendall, W. S. (1986b) The Brownian coupling property and nonnegative Ricci curvature, *Stochastics* **19**, 111–129.

Kendall, W. S. (to appear) The radial part of Brownian motion on a manifold; semimartingale properties. *Ann. Prob.*

Kifer, Yu. (1976) Brownian motion and harmonic functions on manifolds of negative curvature, *Th. Prob. Applic.* **21**, 81–95.

Kifer, Yu. (1982) Entropy via random perturbations, *TAMS* **282**, 589–601.

Kobayashi, S. and Nomizu, K. (1963) *Foundations of Differential Geometry I*, Wiley-Interscience, New York.

Kunita, H. (1984) Stochastic differential equations and stochastic flows of homeomorphisms, in *Stochastic Analysis and Applications*, Advances in Probability and Related Topics No. 7, Marcel Dekker, New York.

Li, P. and Schoen, R. (1984) L^p and mean-value properties of subharmonic functions on Riemannian manifolds, *Acta Math.* **153**, 279–301.

Lindvall, T. and Rogers, L. C. G. (1986) Coupling of multidimensional diffusions by reflection, *Ann. Prob.* **14**, 860–872.

Lyons, T. and McKean, H. P. (1984) Winding of the plane Brownian motion, *Adv. Math.* **51** 212–225.

Lyóns, T. and Sullivan, D. (1984) Function theory, random paths, and covering spaces, *J. Diff. Geom.* **19**, 299–323.

McConnell, J. (1980) *Rotational Brownian Motion and Dielectric Theory*, Academic Press, New York.

McKean, H. P. (1969) *Stochastic Integrals*, Academic Press, New York.

Malliavin, P. (1974) Formule de la moyenne, calcul de perturbation, et théorèmes d'annulation pour les formes harmoniques, *J. Funct. Anal.* **17**, 274–291.

Malliavin, P. (1978) *Géométrie differerentielle stochastique*, Sem. de Math. Sup. Montreal.

Manabe, S. (1982) Stochastic Intersection number and homological behaviors of diffusion processes on Riemannian manifolds, *Osaka J. Math.* **19**, 429–457.

Meyer, P.-A. (1981) Geométrie stochastique sans larmes, *Sem. Prob. XV*, LN Math. **850**, Springer-Verlag, Berlin.

Milnor, J. W. (1963) *Morse Theory*, Princeton University Press.

Molchanov, S. A. (1975) Diffusion Processes and Riemannian Geometry, *Russian Math. Surveys* **30**, 1–53.

Norris, J., Roger, L. C. G., and Williams, D. (1986) Brownian motion of ellipsoids, *TAMS* **294**, 757–765.

Orihara, A. (1970) On random ellipsoid, *J. Fac. Sci. Univ. Tokyo* **17**, 73–85.

Pinsky, M. (1977) An individual ergodic theorem for Brownian motion on a surface of negative curvature, in *Proc. Conf. Stochastic Differential Equations*, Academic Press, New York, pp. 231–240.

Pinsky, M. (1978) Stochastic Riemannian geometry, in Bharucha-Reid (eds.), *Probabilistic Analysis and Related Topics*, **1**, Academic Press, London.

Pinsky, M. (1983) Brownian motion and Riemannian geometry, in A. Gray *et al.* (eds.) *Differential Geometry*, Birkhauser, Boston.

Pitman, J. W. and Yor, M. (1984) The asymptotic joint distribution of windings of planar Brownian motion, *Bull. AMS.* **10**, 109–111.

Prat, J.-J. (1975) Étude asymptotique et convergence angulaire du mouvement brownien sur une variété a courbure negative. *CRAS* **A280**, 1539–1542.

Price, G. C. and Williams, D. (1983) Rolling with slipping I, *Sem. Prob. XVII*, LN Math. **986**, Springer-Verlag, Berlin.

Roberts, P. H. and Ursell, H. D. (1960) Random walk on a sphere and on a Riemannian manifold, *J. Roy. Soc.* **A252**, 317–356.

Schwartz, L. (1984) *Semimartingales and their Stochastic Calculus*, Univ. Montreal.

Spitzer, F. (1964) Electrostatic capacity, heat flow, and Brownian motion, *ZW* **3**, 110–121.

Stroock, D. W. and Varadhan, S. R. S. (1979) *Multidimensional Diffusion Processes*, Springer-Verlag, Berlin.

Sullivan, D. (1983) The Dirichlet problem at infinity for a negatively curved manifold, *J. Diff. Geom.* **18**, 723–732.

Van Den Berg, M. and Lewis, J. T. (1985) Brownian motion on a hypersurface, *Bull. LMS.* **17**, 144–150.

Varadhan, S. R. S. (1967) Diffusion processes in a small time interval, *Comm. Pure Appl. Math.* **20**, 659–685.

Varopoulos, N. Th. (1984) Brownian motion and Random Walks on manifolds, *Ann. Inst. Fourier (Grenoble)* **34** (II), 243–269.

Warner, F. W. (1971) *Foundations of Differentiable Manifolds and Lie Groups*, Scott, Foreman & Co.

Watling, K. (in preparation) Elementary formulae for the heat kernel. Preprint.

Yamada, T. (1973) On a comparison theorem for solutions of stochastic differential equations and its applications, *J. Math. Kyoto Univ.* **13**, 497–512.

Yau, S.-T. (1975) Harmonic functions on complete Riemannian manifolds, *Comm. Pure Appl. Math.* **28**, 201–228.

Yau, S.-T. (1978) On the heat kernel of a complete Riemannian manifold, *J. Math. pures appl.* **57**, 191–201.

Yosida, K. (1949) Brownian motion on the surface of the 3-sphere, *Ann. Math. Statist.* **20**, 292–296.

Zheng, W. A. (1983) Sur le théorème de convergence des martingales dans une variété Riemanniane, *ZW* **63**, 511–515.

Acta Applicandae Mathematicae **9** (1987), 61–69.
© 1987 *by D. Reidel Publishing Company*

Extremal Properties of Some Geometric Processes[*]

J. MECKE
Friedrich-Schiller-Universität, Sektion Mathematik, 6900 Jena, G.D.R.

(Received: 27 July 1986)

Abstract. In this paper some isoperimetric inequalities for stationary random tessellations are discussed. At first, classical results on deterministic tessellations in the Euclidean plane are extended to the case of random tessellations. An isoperimetric inequality for the random Poisson polygon is derived as a consequence of a theorem of Davidson concerning an extremal property of tessellations generated by random lines in R^2. We mention extremal properties of stationary hyperplane tessellations in R^d related to Davidson's result in case $d = 2$. Finally, similar problems for random arrangements of r-flats in R^d are considered ($r < d - 1$).

AMS subject classification (1980). 60D05, 60G55.

Key words. Random tessellations, line processes, flat processes, Poisson polygon, isoperimetric inequalities.

1. Introduction

In several fields of contemporary natural sciences, the question arises of a calculation of quantities which depend on closest packings of balls with a random radius. Problems of this kind turn out to be extremely difficult. Most work done up to the present concerns approximation formulae or experimental results.

The question arises, can stochastic geometry be developed so as to contribute to the solution of these problems?

In support of this, note that in the deterministic case Rogers [12] pointed out that there are relations between closest packings of balls and properties of special tessellations.

In the present paper, it is shown that certain extremal problems for random tessellations can be solved. As an example, stationary random tessellations in the Euclidean plane are treated. Some methods generalize to three-dimensional Euclidean space, but only in the restricted case of random tessellations that are generated by random planes (plane processes). Finally, results for higher-dimensional spaces are mentioned (hyperplane processes and flat processes in R^d).

2. Stationary Random Planar Tessellations

Roughly speaking, a random planar tessellation is a random division of R^2 into bounded convex polygonal regions (Figure 1). Random tessellations are investigated, e.g., in [1, 2, 11, 13].

[*] This work was done while the author was visiting the University of Strathclyde in Glasgow.

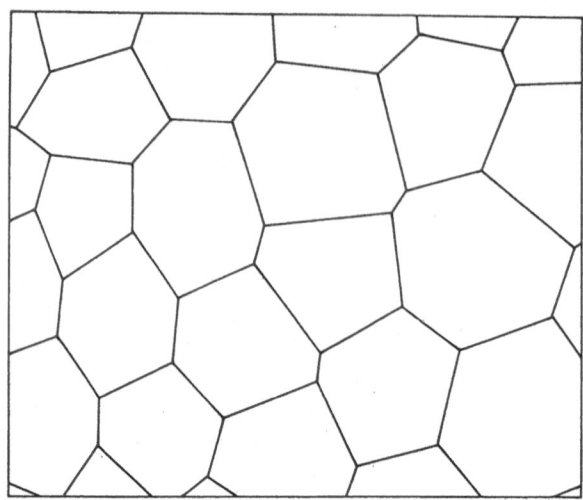

Fig. 1.

We vote for the following definition, cf. [2, 7, 8]. Let $\mathscr{Z}$ be the set of all compact convex polygonal regions in R^2 with a positive area. By $\mathring{Z}$ we denote the open kernel of a polygon $Z \in \mathscr{Z}$. A subclass $m \subset \mathscr{Z}$ is called a planar tessellation if it has the following properties:

(a) $\mathring{Z}_1 \cap \mathring{Z}_2 = \emptyset$ if $Z_1, Z_2 \in m$, $Z_1 \neq Z_2$,
(b) $\bigcup \{Z : Z \in m\} = R^2$,
(c) Every bounded subset $B \subset R^2$ is hit only by a finite number of polygons $Z \in m$.

Let $\mathscr{M}$ be the set of all such tessellations m. By $\mathfrak{M}$ we denote the σ-algebra in $\mathscr{M}$ generated by all sets $\{m \in \mathscr{M} : K_m \cap G = \emptyset\}$, where $K_m = \bigcup \{\partial Z : Z \in m\}$ is the set of all boundary points of the tiles $Z \in m$, and G runs through the system of all open subsets of R^2.

A random planar tessellation is a random variable ϕ with range $[\mathscr{M}, \mathfrak{M}]$. Its distribution is a probability measure P on $[\mathscr{M}, \mathfrak{M}]$.

For every $\mathbf{x} \in R^2$ we define a translation operator

$$T_{\mathbf{x}} : \mathscr{M} \to \mathscr{M} \text{ by } T_{\mathbf{x}} m = \{Z + \mathbf{x} : Z \in m\}.$$

A random tessellation ϕ and its distribution P are said to be stationary if for every $\mathbf{x} \in R^2$ the shifted random tessellation $T_{\mathbf{x}}\phi$ has the same distribution as ϕ, i.e., $P \circ T_{\mathbf{x}}^{-1} = P$.

A simple example may be constructed as follows. Let s be the regular square tessellation with vertices at (k, l), where k, l run through all integers, and let $\xi \in R^2$ be a random vector uniformly distributed in the square

$$\{(x_1, x_2) : 0 \leqslant x_1, x_2 \leqslant 1\}.$$

Then $T_{\xi}s$ is a stationary random tessellation, a so-called stationary square

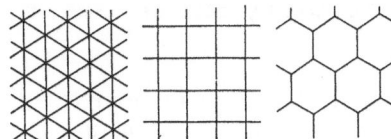

Fig. 2.

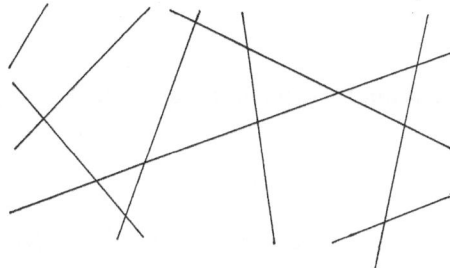

Fig. 3.

tessellation. In a similar way, stationary regular triangle and hexagon tessellations can be defined (Figure 2).

Random tessellations which are generated by random arrangements of lines are called random line tessellations (Figure 3). The most important type is represented by the stationary Poisson line tessellations, they are generated by a stationary Poisson line process [6, 11, 13, 15].

Let ϕ be any stationary random tessellation. The mean total length of boundary lines per unit area is said to be the fibre intensity of ϕ and is usually denoted by λ in this paper. Another mean value of interest is the vertex intensity g, i.e., the mean number of vertices per unit area.

Concerning the definition of other mean values like edge intensity, tile intensity, mean perimeter of the typical tile etc. see, e.g., [7, 8, 15], cf. [2].

3. Problem

In certain classes of nontrivial stationary random planar tessellations, we want to define those for which the fibre intensity is minimal.

In this respect the class of all stationary random tessellations with given vertex intensity is not interesting, because the minimum of the fibre intensity is then zero.

Examples of interesting special classes are, e.g.,

– Stationary random tessellations all tiles of which have the same constant area $F > 0$.
– Stationary random tessellations all tiles of which have the same constant perimeter $U > 0$.
– Stationary Poisson line tessellations with given vertex intensity.

There exist analogous for the last problem in higher dimensions.

4. Stationary Random Planar Tessellations with Constant Area or Constant Perimeter of Tiles

We need the notion of the 'typical tile' of a stationary random tessellation. Roughly speaking, this means a randomly chosen tile, where all tiles have the same chance of being chosen. For a precise definition see [6–8, 15], cf. [2]. Let us denote by n the mean number of neighbouring tiles of the typical tile.

The following results are proved in [8].

THEOREM 4.1. *Suppose all tiles of a stationary random tessellation have constant area $F > 0$. Then*

$$\lambda^2 \geq n \tan(\pi/n)/F.$$

The equality sign holds iff $n = 3, 4, 6$, and all tiles are regular n-sided polygons.

THEOREM 4.2. *Suppose all tiles of a stationary random tessellation have constant perimeter $U > 0$. Then*

$$\lambda \geq 2n \tan(\pi/n)/U.$$

The equality sign holds iff $n = 3, 4, 6$, and all tiles are regular n-sided polygons.

Since the function $n \rightarrow n \tan(\pi/n)$ is decreasing and $n \leq 6$ (cf. [15]), we have

COROLLARY 4.3. *Suppose F (or U) is a positive number. In the class of all stationary random tessellations, all tiles of which have the same area F (perimeter U) the stationary regular hexagon tessellations has the minimal fibre intensity.*

COROLLARY 4.4. *In the class of all stationary random tessellations with $n \leq 4$, all tiles of which have the same area F (perimeter U), the stationary regular square tessellation has the minimal fibre intensity.*

In the case of deterministic tessellations, results of the above kind are well-known, see, e.g., [4]. It is possible to consider the system of notions and conclusions which is developed in [8] and is based on measure theory as a special language to describe and prove the essential content of the classical theory in an elegant manner.

5. Stationary Poisson Line Tessellations

This section is concerned with a very special type of random tessellations, namely those generated by random arrangements of lines. We consider only the case where the random lines are chosen according to a Poisson law.

More precisely, let ϕ be a stationary Poisson line process in the plane, i.e., a translation invariant 'point process' on the space of lines in R^2, see [3, 6, 9, 15].

The fibre intensity of the corresponding tessellation we denote by λ, where $0 < \lambda < \infty$. For a Borel subset $B \subset [0, \pi)$ let ϕ_B be the process of those lines from ϕ, the angle of which with the x_1-axis in the upper half plane belongs to B. If the fibre intensity of ϕ_B is denoted by $\lambda\zeta(B)$, a probability measure ζ on $[0, \pi)$ is defined, the so-called rose of directions of ϕ. The distribution of the Poisson line process ϕ is uniquely determined by the pair (λ, ζ).

The vertex intensity of a stationary Poisson line process with fibre intensity λ and rose of directions ζ is denoted by $g(\lambda, \zeta)$, it is also called the intersection density of ϕ. We have

$$g(\lambda, \zeta) = \frac{\lambda^2}{2} \int \zeta(d\alpha) \int \zeta(d\beta) |\sin(\alpha - \beta)|, \tag{5.1}$$

see [2, 15].

The stationary process ϕ is called isotropic, if its distribution is not only invariant under translations but also under rotations of the plane. In the isotropic case, ζ equals γ, the uniform distribution on $[0, \pi)$:

$$\gamma(d\alpha) = \frac{1}{\pi} d\alpha.$$

We get $g(\lambda, \gamma) = \lambda^2/\pi$.

According to Section 3, we are interested in such stationary Poisson line tessellations with given vertex intensity for which the fibre intensity is minimal. We use an equivalent reformulation of the

PROBLEM. Given $0 < \lambda < \infty$, for which probability distributions ζ on $[0, \pi)$ is the quantity $g(\lambda, \zeta)$ defined by (5.1) maximal? Davidson proved a special result in [3], from which we deduce the following theorem.

THEOREM 5.1. *For all $\lambda > 0$ and all probability measures ζ on $[0, \pi)$ the inequality*

$$g(\lambda, \zeta) \leq g(\lambda, \gamma) \quad (= \lambda^2/\pi)$$

is fulfilled. The equality sign holds iff $\zeta = \gamma$.

This means that in the class of all stationary Poisson line processes with given fibre intensity λ, exactly the isotropic stationary Poisson line process has the maximal intersection density.

Whereas Davidson used Fourier arguments, the same result is derived in [16] by methods of convex geometry. According to Matheron [6] to every pair (λ, ζ) there corresponds a convex region in R^2, the so-called Steiner compact Λ. Up to a constant, the perimeter of Λ equals λ and the area of Λ equals g. In the isotropic case $(\zeta = \gamma)$, Λ is a circle. Theorem 5.1 corresponds, then, to the classical isoperimetric inequality.

6. The Poisson Polygon

The typical tile in a stationary Poisson line tessellation is called the Poisson polygon. The mean area a and the mean perimeter l of the Poisson polygon are functions of the fibre intensity λ and the vertex intensity g:

$$a = 1/g \tag{6.1}$$

$$l = 2\lambda/g. \tag{6.2}$$

This is an easy consequence of the theory of mean values for stationary random tessellations, presented, e.g., in [15].

From (6.1) and (6.2) we get $l^2 = 4a\lambda^2/g$. Together with the Davidson theorem 5.1 ($g \leqslant \lambda^2/\pi$) this yields

THEOREM 6.1. *The mean area a and the mean perimeter l of the Poisson polygon fulfil*

$$l^2 \geqslant 4\pi a. \tag{6.3}$$

The equality sign holds exactly in the isotropic case.

If A is the area and P the perimeter of any deterministic bounded convex region in R^2, then it is well-known that $P^2 \geqslant 4\pi A$. Note the remarkable conformity of this so-called isoperimetric inequality with (6.3).

7. Stationary Poisson Hyperplane Processes

At first we are interested in stationary Poisson plane processes in R^3, i.e., Poisson 'point processes' ϕ on the space of all planes in R^3, the distribution of which is invariant under translations. They generate three-dimensional tessellations.

The mean total area of plane pieces from ϕ per unit volume is said to be the intensity of ϕ and is denoted by λ.

Let $\mathscr{X}$ be the set of planes through the origin. To every plane p in R^3 there corresponds a unique plane $\tau p \in \mathscr{X}$ which is parallel to p. We say that τp indicates the direction of p. The rose of directions ζ of ϕ is a probability measure on $\mathscr{X}$, it describes the distribution of directions of the planes in ϕ and is defined in an analogous manner as in the case of line processes in Section 5. In the isotropic case, ζ equals γ, the unique rotation invariant probability measure on $\mathscr{X}$.

The set of intersection lines of plane pairs taken from ϕ is said to be the one-dimensional intersection manifold and denoted by $s_1(\phi)$. This is, in fact, a stationary line process in R^3. The mean total length of line pieces from $s_1(\phi)$ per unit volume is called one-intersection density. It depends on the intensity λ and the rose of directions, hence we denote it by $i_1(\lambda, \zeta)$. From the results of Janson and Kallenberg in [5] the following statement can be derived.

THEOREM 7.1. *For all $\lambda > 0$ and all probability measures ζ on $\mathscr{X}$, we have $i_1(\lambda, \zeta) \leqslant i_1(\lambda, \gamma)$. The equality sign holds iff $\zeta = \gamma$.*

This means that under all stationary Poisson plane processes in R^3 with given intensity $\lambda > 0$, exactly the isotropic process has the maximal one-intersection density.

Parallel to the one-dimensional intersection manifold we define a zero-dimensional intersection manifold $s_0(\phi)$ consisting of all intersections points determined by the plane triples taken from ϕ. It is, in fact, a stationary point process in R^3. Its intensity (mean number of points per unit volume) is called zero-intersection density and denoted by $i_0(\lambda, \zeta)$. The results of Thomas in [16] imply the following theorem.

THEOREM 7.2. *For all $\lambda > 0$ and all probability measures ζ on $\mathfrak{X}$ we have $i_0(\lambda, \zeta) \leqslant i_0(\lambda, \gamma)$. The equality sign holds iff $\zeta = \gamma$.*

This means that under all stationary Poisson plane processes in R^3 with given intensity $\lambda > 0$, exactly the isotropic process has the maximal zero-intersection density.

Following Davidson, Janson and Kallenberg used Fourier methods to get their results. But proofs of this kind for Theorem 7.2 seem to be unknown. As outlined in Section 5, in the two-dimensional case, Thomas has exploited isoperimetric properties of the Steiner compact. This is, in our present three-dimensional case, a convex body Λ in R^3 which is assigned to the pair (λ, ζ) in a unique manner, see Matheron [6]. In the isotropic case, Λ is a ball. Up to a constant, the volume of Λ equals i_0, the surface area equals i_1, and the average breadth equals λ. Theorems 7.1 and 7.2 are then consequences of classical isoperimetric inequalities for convex bodies in R^3.

In this way Thomas has proved analogous results for stationary Poisson hyperplane processes in R^d [16]. With respect to these processes, one can regard the intensities i_k of k-dimensional intersection manifolds $(k = 0, 1, \ldots, d-2)$. The intensity of the process itself is again denoted by λ, i.e., the mean $(d-1)$-content of hyperplane pieces per unit d-volume. It turns out that for given intensity λ and each $k \in \{0, 1, \ldots, d-2\}$ the k-intersection density of a stationary Poisson hyperplane process in R^d is maximal exactly in the isotropic case.

These results may also be deduced from integral geometric theorems of Schneider [14], which, finally, are also based on isoperimetric inequalities for convex bodies in R^d.

8. Stationary Poisson Flat Processes

An r-flat in R^d is an r-dimensional linear manifold (affine subspace) in R^d. The hyperplanes considered in Section 7 are $(d-1)$-flats. We are now interested in stationary Poisson r-flat processes $(d/2 \leqslant r < d-1)$ in R^d, i.e., in stationary Poisson 'point processes' on the phase space of all r-flats in R^d. Due to the assumption $r \geqslant d/2$ there are nonempty intersection manifolds. But the extremal properties are quite different from those discussed in Section 7 $(r = d-1)$.

As an example, we regard the special case $d = 4$, $r = 2$. For fixed $\lambda > 0$ let ϕ_j and ϕ_s be stationary Poisson two-flat processes in R^4 with mean two-content λ per unit four-volume (intensity λ). The process ϕ_j should be isotropic, whereas ϕ_s should be the superposition of two independent flat processes ϕ_{p1}, ϕ_{p2}, both of them with mean two-content $\lambda/2$ per unit four-volume. Each process ϕ_{p1}, ϕ_{p2} only contains flats which are parallel to each other, and the flats of ϕ_{p1} are perpendicular to those of ϕ_{p2}. Almost every flat pair in ϕ_j determines a unique intersection point, also does every flat pair in ϕ_s. Hence, the intersection manifolds of both processes are stationary point processes in R^4, the corresponding intensities are denoted by i_j and i_s. We find $i_j = \lambda^2/6$, $i_s = \lambda^2/4$ and, consequently, $i_s > i_j$. This means that in contrast to the situation in Section 7, here the isotropic process is not extremal.

Let now ϕ be a stationary Poisson r-flat process in R^d ($d \geq 4$, $d/2 \leq r < d - 1$). The mean r-content of r-flats per unit d-volume is called intensity of ϕ and denoted by λ. The set of intersections of flat pairs taken from ϕ is said to be the intersection manifold (of second order) of ϕ, it is a stationary $(2r-d)$-flat process (non-Poisson). Its intensity ($(2r - d)$-content per unit d-volume) is called intersection density (of second order).

The results in [10] imply the following theorem.

THEOREM 8.1. *For every $d \geq 4$, $d/2 \leq r < d - 1$ and $\lambda > 0$ there exists a stationary Poisson r-flat process in R^d with intensity λ the intersection density (of 2nd order) of which is greater than that of the isotropic stationary Poisson r-flat process with intensity λ.*

In [10] the class of extremal processes is not determined. Moreover, the extremal properties concerning intersection manifolds of higher order (intersections of more than two flats of the process) are not treated.

References

1. Ambartzumian, R. V.: 'Convex Polygons and Random Tessellations', in E. F. Harding and D. G. Kendall (eds.), *Stochastic Geometry*, Wiley, New York, 1974.
2. Cowan, R.: 'The Use of Ergodic Theorems in Random Geometry', *Suppl. Adv. Appl. Prob.* **10** (1978), 47–57.
3. Davidson, R.: 'Line Processes, Roads and Fibres', in E. F. Harding and D. G. Kendall (eds.), *Stochastic Geometry*, Wiley, New York, 1974.
4. Fejes Toth, L.: *Reguläre Figuren*, Teubner, Leipzig, 1965.
5. Janson, S. and Kallenberg, O.: 'Maximizing the Intersection Density of Fibre Processes', *J. Appl. Prob.* **18** (1981), 820–828.
6. Matheron, G.: *Random Sets and Integral Geometry*, Wiley, New York, 1975.
7. Mecke, J.: 'Palm Methods for Stationary Random Mosaics', in R. V. Ambartzumian (ed.), *Combinatorial Principles in Stochastic Geometry*, Armenian Acad. of Sc. Publishing House, Erevan, 1980.
8. Mecke, J.: 'Isoperimetric Properties of Stationary Random Mosaics', *Math. Nachr.* **117** (1984), 75–82.
9. Mecke, J.: 'On Some Inequalities for Poisson Networks', *Math. Nachr.* **128** (1986), 81–86.

10. Mecke, J. and Thomas, C.: 'On an Extreme Value Problem for Flat Processes', *Commun. Statist. – Stochastic Models* **2**(2) (1986), 273–280.
11. Miles, R. E.: 'A Synopsis of Poisson Flats in Euclidean Spaces', in E. F. Harding and D. G. Kendall (eds.), *Stochastic Geometry*, Wiley, New York, 1974.
12. Rogers, C. A.: *Packing and Covering*, Cambridge Tracts in Mathematics and Mathematical Physics, No. 54, 1964.
13. Santaló, L. A.: *Integral Geometry and Geometric Probability*, Addison-Wesley, London, 1976.
14. Schneider, R.: 'Random Hyperplanes Meeting a Convex Body', *Z. Wahrscheinlichkeitstheorie verw. Geb.* **61** (1982), 379–387.
15. Stoyan, D., Kendall, W., and Mecke, J.: *Stochastic Geometry*, Wiley, New York; Akademie Verlag, Berlin, 1987.
16. Thomas, C.: 'Extremum Properties of the Intersection Densities of Stationary Poisson Hyperplane Processes', *Math. Opforsch. Statist., ser. statistics,* **15** (1984), 443–449.

Note added in proof. The deeper reason for the conformity of (6.3) with the usual isoperimetric inequality is the following:

The mathematical expectation of the supporting function of the Poisson polygon is the supporting function of a convex region which may be called the mean Poisson polygon. It is easy to see that the mean perimeter l of the Poisson polygon is equal to the perimeter of the mean Poisson polygon. But it is also true that the mean area a of the Poisson polygon is equal to the area of the mean Poisson polygon. Hence, the inequality (6.3) corresponds to the isoperimetric inequality for the (deterministic) mean Poisson polygon.

Acta Applicandae Mathematicae **9** (1987), 71–81.
© 1987 by D. Reidel Publishing Company.

Combinatorial Decompositions and Homogeneous Geometrical Processes

V. K. OGANIAN

Department of Mathematics, Yerevan State University, Mravian Street 1, Yerevan 375049, Armenia, U.S.S.R.

(Received: 27 June 1986)

Abstract. This paper considers line processes and random mosaics. The processes are assumed invariant with respect to the group of translations of $\mathbf{R}^2$. An expression for the probabilities $\pi_k(t, \alpha)$, $k = 0, 1, 2, \ldots$ to have k hits on an interval of length t taken on a 'typical line of direction α' (the hits are produced by other lines of the process) is obtained. Also, the distribution of a length of a 'typical edge having direction α' in terms of the process $\{\mathcal{P}_i, \psi_i\}$ is found, here $\mathcal{P}_i$ is the point process of intersections of edges of the mosaic with a fixed line of direction α and the mark ψ_i is the intersection angle at $\mathcal{P}_i$. The method is based on the results of combinatorial integral geometry.

AMS subject classifications (1980). 60D05, 60G55.

Key words. Line processes, random mosaics, combinatorial decompositions, marked point processes.

1. Introduction

The method of combinatorial decompositions was used by the author in his study of random geometrical processes in [4–7] (see also [1–3,9]). The results concerned the stereology of random processes of lines, segments (in particular, random mosaics) or Boolean models in the plane under assumption of their invariance with respect to the group $\mathbf{M}_2$ of Euclidean motions of $\mathbf{R}^2$.

Our aim here is to obtain similar results for planar random geometrical processes which are invariant with respect to the group $\mathbf{T}_2$ of translations of $\mathbf{R}^2$. Because of space limitations, we consider here only line processes and random mosaics although Boolean models yield the same type of analysis.

The method is based on the results of combinatorial integral geometry (see the book of R. V. Ambartzumian [1] for a complete account of this theory). We recall here the main combinatorial decompositions for the space $\mathbf{G}$ of lines in the plane.

Let $\{\mathcal{P}_i\}_{i=1}^n$ be a fixed finite set of points in $\mathbf{R}^2$ with no three points on the same line. Denote by ν_{ij} the segment with endpoints $\mathcal{P}_i$ and $\mathcal{P}_j$ and $[\nu_{ij}] = \{g \in \mathbf{G}: g \cap \nu_{ij} \neq \phi\}$. Let $a\{\mathcal{P}_i\}$ be the minimal (finite) algebra subsets of $\mathbf{G}$ which contains all $[\nu_{ij}]$ and $\mathrm{Br}\{\mathcal{P}_i\}$ be the ring of bounded members of $a\{\mathcal{P}_i\}$.

For each $B \in \mathrm{Br}\{\mathcal{P}_i\}$ its $\mathbf{M}_2$-invariant measure μ can be represented in the form

$$\mu(B) = \sum_{i<j} c_{ij}(B) \cdot |\nu_{ij}|, \tag{1}$$

where $|\nu|$ stands for the length, the sum is extended over the set of all non-ordered pairs $\mathscr{P}_i$, $\mathscr{P}_j$, and $c_{ij}(B)$ are integers with possible values $0, \pm 1, \pm 2$. An algorithm of calculation of these coefficients is given in [1] and [8].

The method consists of averaging formulae of the type (1) written for parts of realizations within a narrow rectangle. We consider two main types of geometrical processes (Poissonity is not assumed):

(a) line processes in Section 4,

(b) random mosaics in Section 5.

In Section 4 we obtain an expression for the probabilities $\pi_k(t, \alpha)$, $k = 0, 1, 2, \ldots$ to have k hits on an interval of length t taken on a 'typical line of direction α' (the hits are produced by other lines of the process).

In Section 5 we find the distribution of length of a 'typical edge having direction α' in terms of the process $\{\mathscr{P}_i, \psi_i\}$. Here $\{\mathscr{P}_i\}$ is the point process of intersections of the edges of the mosaic with a fixed line of direction α and the mark ψ_i is the intersection angle at $\mathscr{P}_i$.

2. Description of the Main Tool

Let us give a short description of the main idea. Let ν_1, ν_2 be two parallel segments of the length l distance t apart situated as sides of a rectangle R (see Figure 1).

Let $m(\omega)$ be a $\mathbf{T}_2$-invariant random geometrical process in $\mathbf{R}^2$ of one of the types (a) and (b) mentioned above.

By $\mathfrak{N}(\omega)$ we denote the following set: in the case (a) $\mathfrak{N}(\omega)$ is a union of all lines which belong to the realization in Section 4; in the case (b) $\mathfrak{N}(\omega)$ is a union all edges of the mosaic in Section 5.

For any fixed realization $m(\omega_0)$ we consider $\mathfrak{N}(R) = \mathfrak{N}(\omega_0) \cap R$ (the trace of $\mathfrak{N}(\omega)$ in R). We average the invariant measure $\mu(B_k \cap C)$ (where

$$B = \{g \in \mathbf{G} : \operatorname{card}(\mathfrak{N}(R) \cap g) = k, \quad k = 0, 1, 2, \ldots$$

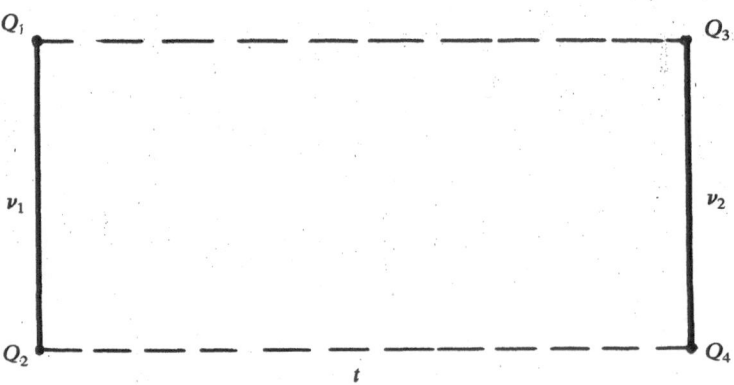

Fig. 1. The segment $Q_2 Q_4$ of length t has the direction α.

and $C = [\nu_1] \cap [\nu_2])$ with respect to the T_2-invariant distribution P of the random geometrical processes in question.

Using the Fubini theorem we get:

$$E_P \mu(B_k \cap C) = \int_C p_k(x(g), \varphi) \, dg. \tag{2}$$

Here the function $\text{card}(\mathfrak{N}(R) \cap g)$ is the number of points of the set $\mathfrak{N}(R) \cap g$; E_P denotes expectation with respect to distribution P of $m(\omega)$; dg is the element of the invariant measure $\mu(dg) = dp \, d\varphi$, where p is the distance from the origin 0 to g and φ is the direction of g, see [8]), and

$$p_k(x, \varphi) = P\binom{\chi}{k}$$

where χ is a line segment of the length x and the direction φ,

$$\binom{\chi}{k} = \left\{ \begin{matrix} \text{in the segment } \chi \text{ is exactly } k \text{ points} \\ \text{of the set } \mathfrak{N}(R) \end{matrix} \right\}.$$

Assuming that l tends to zero we separate the main term in (2):

$$E_P \mu(B_k \cap C) = \frac{p_k(t, \alpha)}{t} l^2 + o(l^2). \tag{3}$$

Thus, integration of the right side of (1) written for $\mu(B_k \cap C)$ and separation of the terms which are l by order will lead to some representation for $p_k(t, \alpha)$ in terms of the first- and second-order Palm distributions of the event $\binom{\chi}{k}$.

3. Necessary Notions and Notations

By integrating the right-hand side of combinatorial formula (1) written for $\mu(B_k \cap C)$, we will use the following two formulae which are well known in the theory of point processes. They connect the expectations of random sums with the Palm distributions of the processes (see [9, 11, 3]):

$$E_P \sum_{x_i \in m} \varphi_1(x_i, m) = \int_X d\Lambda_i(x) \int_{\mathfrak{M}_x} \varphi_1(x, m) \Pi_x(dm), \tag{4}$$

$$E_P \sum_{x_i, x_j \in m} \varphi_2(x_i, x_j, m) = \int_{X \times X} \Lambda_2(dx_1, dx_2) \int_{\mathfrak{M}_x} \varphi_2(x_1, x_2, m) \, \Pi_{x_1, x_2}(dm). \tag{5}$$

Here $\varphi_1 : X \times \mathfrak{M}_x \to R^1$ and $\varphi_2 : X \times X \times \mathfrak{M}_x \to R^1$ are measurable, $\mathfrak{M}_x$ is the space of the realizations of a point process $m(\omega)$ on a manifold X, $\Lambda_1(\cdot)$ and $\Lambda_2(\cdot)$ are the first and second moment measures of the point process; $\Pi_x(\cdot)$ and $\Pi_{x_1, x_2}(\cdot)$ are the first- and second-order Palm distributions of the point process. In fact, the identities (4) and (5) can serve as definition for Π_x, Π_{x_1,x_2} respectively; we note, however, that Π_x and Π_{x_1,x_2} have intuitive interpretations of conditional distributions of the point process $m(\omega)$ under the condition that at $x \in \mathbf{X}$ (or in $(x_1, x_2) \in \mathbf{X} \times \mathbf{X}$) there are points from $m(\omega)$.

In all cases we suppose that $m(\omega)$ is of the second order, i.e., both measures Λ_1 and Λ_2 are local finite.

4. Random Processes of Lines

Random processes of lines in the plane are usually defined to be point processes on a manifold $\mathbf{G}$, representing the space of lines $g \in \mathbf{G}$ (see, for example, the introduction in [8] and [10]).

The distribution of $m(\omega)$ is a probability P on $\mathfrak{M}_\mathbf{G}$ (or rather on an appropriate σ-algebra on $\mathfrak{M}_\mathbf{G}$). The group $\mathbf{T}_2$ of all translations of $\mathbf{R}^2$ induces a transformation group of $\mathfrak{M}_\mathbf{G}$. An $m(\omega)$ is called homogeneous ($\mathbf{T}_2$-invariant) if its distribution P is invariant with respect to this group.

The best known $\mathbf{T}_2$-invariant processes of lines are those of Poisson governed by the measure $\lambda\, dp\, m(d\varphi)$ where dp is the element of Lebesgue measure on $\mathbf{R}^1$ and $m(\cdot)$ is a measure on circle $\mathbf{S}_1$. The first moment measure $\Lambda_1(\cdot)$ for *any* $\mathbf{T}_2$-invariant random process of lines necessarily has the factorized form

$$\lambda\, dp\, m(d\varphi).$$

Below we assume that the measure $m(\cdot)$ has a continuous density, i.e., $\Lambda_1(\cdot)$ has a form:

$$\lambda f(\varphi)\, dp\, d\varphi \tag{6}$$

where $\lambda > 0$ is a constant.

We will assume also that the second moment measure $\Lambda_2(\cdot)$ is absolutely continuous with respect to $dg_1\, dg_2$:

$$\Lambda_2(dg_1\, dg_2) = f(g_1, g_2)\, dg_1\, dg_2 \tag{7}$$

where the density $f(g_1, g_2)$ is assumed continuous.

Let us write the combinatorial formula (1) for $\mu(B_k \cap C)$. Now the set of points generating the Buffon ring is $\{\mathscr{P}_i\} \cup \{Q_j\}_{j=1}^4$ where $\{\mathscr{P}_i\}$ are the endpoints of segments $\{\chi_i\} = \mathfrak{N}(R)$ (see Figure 2) and $\{Q_j\}_{j=1}^4$ the set of endpoints of segments ν_1 and ν_2 (see Figure 1). The result has a form

$$\mu(B_k \cap C) = \sum_{i=1}^4 A_i \tag{8}$$

where

$$A_1 = \rho_{14} I_k(\rho_{14}) + \rho_{23} \cdot I_k(\rho_{23}) - t \cdot I_k(\rho_{13}) - t \cdot I_k(\rho_{24}), \tag{9}$$

$$A_2 = -2 \sum_i |\chi_i| \cdot I_C(\chi_i) \cdot \Delta I_{k-1}(\chi_i), \tag{10}$$

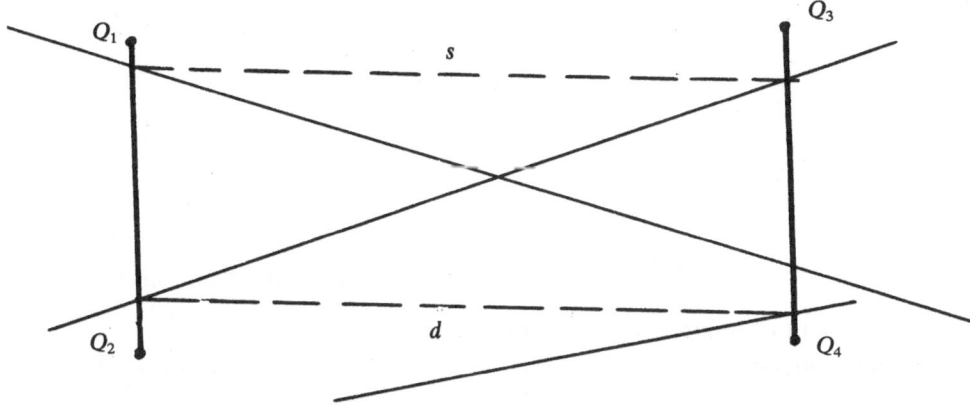

Fig. 2. The solid lines show the segments χ_i. The dotted lines join the pairs of endpoints of the χ_i segments. They can be of d or s type. The type depends on the directions of the two solid segments which proceed from the endpoints of the dotted segments. The type is s if both latter segments lie in one halfplane with respect to the dotted segments, d otherwise.

$$A_3 = \sum_i |\gamma_i| \cdot I_C(\gamma_i) \cdot \Delta^2 I_{k-2}(\gamma_i) \cdot [I_d - I_s], \tag{11}$$

$$A_4 = -\sum_i |\beta_i| \cdot I_C(\beta_i) \cdot \Delta I_{k-1}(\beta_i) \cdot [I_d - I_s]. \tag{12}$$

Here ρ_{14} and ρ_{23} are the diagonals of rectangle R, ρ_{13} and ρ_{24} are its longer sides; $I_k(\tau) = I_{B_k}(\tau) = I$, if the line g_τ on which τ lies belongs to B_k;

$$I_C(\tau) = I, \quad \text{if } g_\tau \in C;$$
$$\Delta I_{k-1} = I_k - I_{k-1}; \quad \Delta^2 I_{k-2} = I_k - 2 \cdot I_{k-1} + I_{k-2};$$
$$\Delta I_{-1} = \Delta^2 I_{-2} = I_0; \quad \Delta^2 I_{-1} = I_0 - 2 \cdot I_1;$$

Segments of β type are the segments connecting a point Q_j ($j = 1,2,3,4$) with a point $\mathscr{P}_i$ and γ is a segment with endpoints from $\{\mathscr{P}_i\}$ which does not belong to the set $\{\chi_i\}$.

Other notations are shown on Figure 2.

So let P be a $\mathbf{T}_2$-invariant probability on $\mathfrak{M}_G$ which satisfies the conditions (6) and (7) and is concentrated on the realizations which have no parallel lines.

Integrating (9) with respect to P we get

$$E_P A_1 = \sqrt{t^2 + l^2}(p_k(\sqrt{t^2 + l^2}, \alpha + \varphi) + p_k(\sqrt{t^2 + l^2}, \alpha - \varphi)) - 2tp_k(t, \alpha),$$

where $\varphi = \arctan(l/t)$.

Separation of the main term as $l \to 0$ yields

$$E_P A_1 = \left(\frac{\partial p_k(t, \alpha)}{\partial t} + \frac{p_k(t, \alpha)}{t} + \frac{\partial^2 p_k(t, \alpha)}{\partial \alpha^2} \cdot \frac{1}{t} \right) \cdot l^2 + o(l^2). \tag{13}$$

Using formula (4) and then (6), we find

$$E_P A_2 = -2E_P \sum_i |\chi_i| \cdot I_C(\chi_i) \cdot \Delta I_{k-1}$$

$$= -2\lambda \int_C \chi(g) \cdot \Delta \pi_{k-1}(\chi(g), \varphi) f(\varphi) \, dp \, d\varphi,$$

where $\pi_k(\chi, \varphi) = \Pi_g\binom{\chi}{k}, \binom{\chi}{k}$ is an event that there are k intersections with $m(\omega)$ on the segment of length $\chi(g)$ and of direction φ.

It is not difficult to separate the main term of $E_P A_2$ for $l \to 0$. We obtain

$$E_P A_2 = -2\lambda f(\alpha) \Delta \pi_{k-1}(t, \alpha) l^2 + o(l^2). \tag{14}$$

In a similar way using formulae (5) and (7), we get

$$E_P A_3 = t \cdot \int_0^\pi \int_0^\pi \Delta^2 \Pi_{g_1 g_2}\binom{t}{k-2} \cdot$$
$$\cdot [I_d - I_s] \cdot f(t, \alpha, \psi_1, \psi_2) \cdot \sin \psi_1 \sin \psi_2 \, d\psi_1 \, d\psi_2 \cdot l^2 + o(l^2), \tag{15}$$

where $f(t, \alpha, \psi_1, \psi_2)$ is the density of the second moment measure (see (7)).

Let us consider the last term in (8).

$$A_4 = A_{41} + A_{42} + A_{43} + A_{44},$$

where

$$A_{41} = -\sum_i \psi(\beta_i) \cdot [I_{01} + I_{10}], \tag{16}$$

$$A_{42} = -\sum_i \psi(\beta_i) \cdot I_{110}(m), \tag{17}$$

$$A_{44} = -\sum_i \psi(\beta_i) \cdot \sum_{\substack{k_1, k_2 \\ k_1 + k_2 > 2}} I_{k_1 k_2}(m) \tag{18}$$

$$A_{43} = -\sum_i \psi(\beta_i) \cdot I_{11p}(m), \tag{19}$$

and

$$\psi(\beta) = |\beta| \cdot I_C(\beta) \cdot \Delta I_{k-1}(\beta) \cdot [I_d - I_s],$$
$$I_{k_1 k_2}(m) = I, \quad \text{if } \operatorname{card}(m \cap v_i) = k_i, \quad i = 1, 2,$$
$$I_{110}(m) = I, \quad \text{if } \operatorname{card}(m \cap v_i) = I, \quad i = 1, 2$$

and the segments v_1 and v_2 are intersected by the same line of realization;

$$I_{11P}(m) = I - I_{110}(m);$$

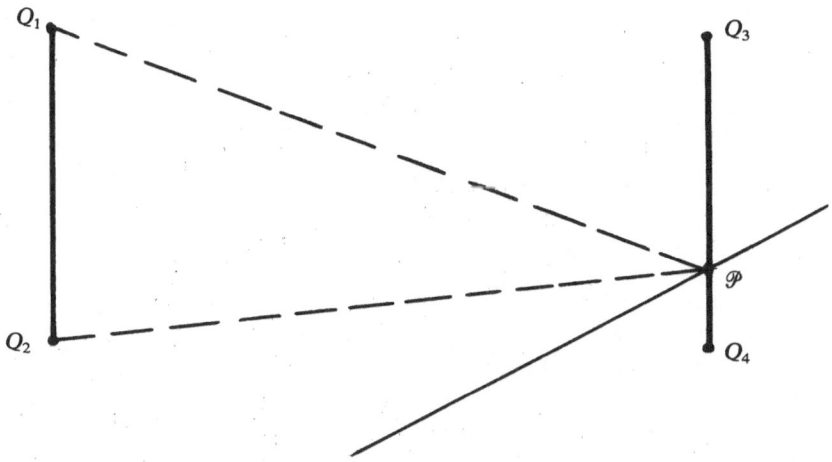

Fig. 3.

Let us show that $E_P A_{41} = o(l^2)$. If, for example, $I_{01}(m) = I$, then $I_k(Q_1, \mathcal{P}) = I_k(Q_2, \mathcal{P})$ for all k (see Figure 3).

Therefore, $I_k(Q_1, \mathcal{P})(|Q_1, \mathcal{P}| - |Q_2, \mathcal{P}|) = o(l^2)$ as $|Q_1, \mathcal{P}| - |Q_2, \mathcal{P}| = O(l^2)$. The case $I_{10}(m) = 1$ can be treated similarly.

It follows that

$$E_P A_{41} = o(l^2), \tag{20}$$

because $E_P(I_{10} + I_{01}) = O(l)$.

It is not difficult to show that

$$E_P A_{44} = o(l^2). \tag{21}$$

We now turn to the term A_{42}. If $I_{110}(m) = I$ (see Figure 4), then for each of the segments $Q_1, \mathcal{P}_{j+1}, Q_2, \mathcal{P}_{j+1}, Q_3, \mathcal{P}_j, Q_4, \mathcal{P}_j, I_s = 1$ and consequently

$$E_P A_{42} = 4\lambda f(\alpha) \cdot \Delta \pi_{k-1}(t, \alpha) l^2 + o(l^2). \tag{22}$$

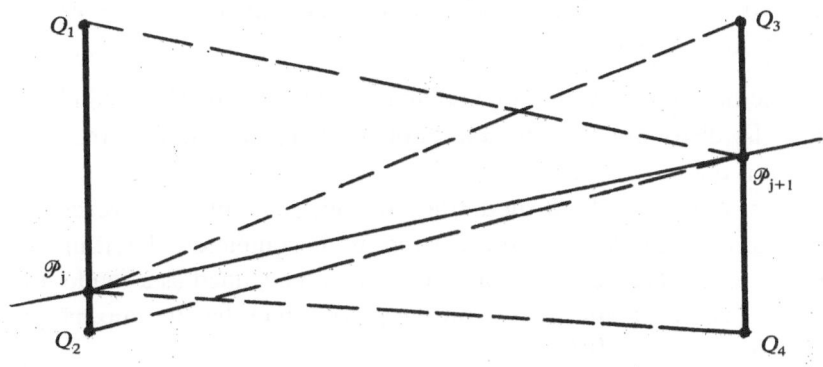

Fig. 4.

Similarly

$$E_P A_{43} = -2 \cdot E_P A_3 + o(l^2). \tag{23}$$

Equating to coefficients of order l^2 in the right and left sides of the averaged formula (8) from (3), (13)–(15) and (20)–(23), we get the main result of this section

$$\frac{\partial p_k(t, \alpha)}{\partial t} = -\frac{1}{t} \frac{\partial^2 p_k(t, \alpha)}{\partial \alpha^2} - 2\lambda f(\alpha) \Delta \pi_{k-1}(t, \alpha) +$$

$$+ t \int_0^\pi \int_0^\pi \Delta^2 \Pi_{g_1 g_2} \left(\frac{t}{k-2} \right) \times$$

$$\times [I_d - I_s] \cdot f(t, \alpha, \psi_1, \psi_2) \sin \psi_1 \cdot \sin \psi_2 \, d\psi_1 \, d\psi_2. \tag{24}$$

This relation can be useful in the study of general T_2-invariant random processes of lines. Under supplementary assumptions of isotropy of the process of lines it was used in [5].

The case where $\pi_k(t, \alpha) = p_k(t, \alpha)$ can be the most convenient to study. We stress that this is a general condition of the mixing type.

If we also assume that the variables $\cot \psi_i$ are noncorrelated (ψ_i is the sequence of intersection angles of lines of the process with axis of direction α), then (24) be reduces to the following differential equation

$$\frac{\partial p_k(t, \alpha)}{\partial t} = -\frac{1}{t} \frac{\partial^2 p_k(t, \alpha)}{\partial \alpha^2} - 2\lambda f(\alpha) \cdot \Delta p_{k-1}(t, \alpha). \tag{25}$$

A problem arises: describe probabilistic solutions of (25) which satisfy the initial conditions:

$$p_k(t, \alpha) = \begin{cases} I, & \text{if } k = 0 \\ 0, & \text{if } k > 0 \end{cases} \quad \text{and} \quad p_k(t, \alpha + \pi) = p_k(t, \alpha).$$

(In the isotropic case, the solution is the family of Poisson distributions.)

5. The Distribution of the Length of the 'Typical Edge of Direction α' in a Random Mosaic

The same method of integration of a combinatorial decomposition can be applied for finding the distribution of the length of the 'typical edge of direction α' for T_2-invariant random mosaic.

A mosaic is defined to be a subdivision of the plane into nonoverlapping and nonempty convex bounded polygons. A mosaic is completely determined by the set of its edges $\{S_i\}$. Hence, a random mosaic can be defined as a random process of edges, i.e., as a segment process on the plane which has the mosaic property with probability one (see [8]).

Denote by $\mathfrak{M}$ the space of mosaics. A probability P on $\mathfrak{M}$ is called the distribution of a random mosaic. The vertices of the polygons from $m(\in \mathfrak{M})$ are called nodes of the mosaic m. A node Q is called a T-type node, if a line through Q exists which leaves all edges starting from Q in one of its closed half-plane. The group $\mathbf{T}_2$ of translations of the plane induces a group acting on $\mathfrak{M}$. A random mosaic is called homogeneous, if its distribution P is invariant with respect to this group ($\mathbf{T}_2$-invariant).

We will suppose that the mean square of the number of edges of mosaic intersecting a bounded Borel set is finite and also with probability one a mosaic has no T-type edges.

A segment in the plane is defined by a quadruple

$$\delta = (p, \varphi, t, \tau) = (g, t, \tau),$$

where (p, φ) are coordinates of the line g which carries the segment, t is one-dimensional coordinate of the center of the segment on g, and τ is the length of δ.

The first moment measure $\Lambda_1(\cdot)$ of the process of edges of a $\mathbf{T}_2$-invariant random mosaic is a measure in the corresponding four-dimensional space. If we assume that $\Lambda_1(\cdot)$ has a density with respect to $d\delta = dp \ d\varphi \ dt \ d\tau$, then it necessarily has a form

$$\lambda f(\tau, \varphi) \, dp \, d\varphi \, dt \, d\tau = \lambda f(\tau, \varphi) \, d\delta. \tag{26}$$

We will assume that the density f is continuous. As for $\Lambda_2(\cdot)$ we will assume that

$$\int_{\Delta_1} \Lambda_2(d\delta_1 \, d\delta_2) = f_1(g_1, g_2) \, dg_1 \, dg_2 \tag{27}$$

with continuous f. Integration in (27) is by the variables $(t_1, t_2, \tau_1, \tau_2)$ over the domain

$$\Delta_1 = \{(t_1, t_2, \tau_1, \tau_2) : \delta_1 \text{ hits } \nu_1, \delta_2 \text{ hits } \nu_2,$$
$$\text{with fixed carrying lines } g_1 \in [\nu_1], g_2 \in [\nu_2]\}.$$

Denote by

$$B_0 = \{g \in \mathbf{G} : R \cap g \text{ intersects no edge of the mosaic}\}.$$

For $\mu(B_0 \cap C)$ we apply the combinatorial decomposition (8) with all terms retaining their form except for the term A_2 which is now

$$A_2 = -2 \sum_i |\chi_i| \cdot I_C(\chi_i).$$

Here χ_i are the chords of R generated by the edges; $\{\mathscr{P}_i\}$ are points of intersection of the edges with ∂R.

Using the formula (4) and then (26) we get:

$$E_P A_2 = -2\lambda \int_t^{+\infty} (\tau - t) f(\tau, t) \, d\tau \cdot l^2 + o(l^2),$$ (28)

where λ is the intensity of the edges.

Expression (13) for $E_P A_1$ remains without change.

In a similar way using formulae (5) and (27) we get (see, also [6]):

$$E_P A_3 = t \cdot \int_0^\pi \int_0^\pi \Pi_{\psi_1, \psi_2, t} \binom{t}{0} [I_d - I_s] \cdot f_1(t, \alpha, \psi_1, \psi_2) \times$$
$$\times \sin \psi_1 \sin \psi_2 \, d\psi_1 \, d\psi_2 \cdot l^2 + o(l^2),$$ (29)

where

$$f_1(t, \alpha, \psi_1, \psi_2) \cdot \Pi_{\psi_1, \psi_2, t} \binom{t}{0} \cdot \sin \psi_1 \cdot \sin \psi_2 \, d\psi_1 \, d\psi_2$$
$$= \lim_{l \to 0} \frac{1}{l^2} \int_{\Delta_1} \Pi_{\delta_1 \delta_2} \binom{t}{0} \Lambda_2(d\delta_1 \, d\delta_2),$$

$f_1(t, \alpha, \psi_1, \psi_2)$ is the density of the second moment measure of the marked point process of intersections $\{\mathscr{P}_i, \psi_i\}$ in a test line of direction α (see the Introduction); $\Pi_{\psi_1, \psi_2, t} \binom{t}{0}$ is a conditional probability of the event $\binom{t}{0}$ under the condition that $|\mathscr{P}_1, \mathscr{P}_2| = t$ and the marks in the points $\mathscr{P}_1$ and $\mathscr{P}_2$ are equal to ψ_1 and ψ_2, respectively.

Repeating the reasoning similar to that of Section 3 we can get

$$E_P A_4 = -2[E_P A_2 + E_P A_3] \cdot l^2 + o(l^2).$$ (30)

Equating the coefficients of order l^2 on the right and left sides of the averaged formula (8) from (3), (13), (28)–(30) we obtain

$$\frac{\partial p_0(t, \alpha)}{\partial t} = -\frac{1}{t} \frac{\partial^2 p_0(t, \alpha)}{\partial \alpha^2} - 2\lambda \int_t^{+\infty} (\tau - t) f(\tau, \alpha) \, d\tau +$$
$$+ t \int_0^\pi \int_0^\pi \Pi_{\psi_1, \psi_2, t} \binom{t}{0} \cdot [I_d - I_s] \times$$
$$\times f_1(t, \alpha, \psi_1, \psi_2) \cdot \sin \psi_1 \cdot \sin \psi_2 \, d\psi_1 \, d\psi_2.$$ (31)

By differentiating (31) with respect to t, we get the main result of this section:

$$P(|\tau| > t, \alpha)$$
$$= \frac{1}{2\lambda} \frac{\partial^2 p_0(t, \alpha)}{\partial t^2} + \frac{1}{2\lambda} \frac{\partial}{\partial t} \left[\frac{1}{t} \cdot \frac{\partial^2 p_0(t, \alpha)}{\partial \alpha^2} \right] -$$
$$- t \int_0^\pi \int_0^\pi \Pi_{\psi_1, \psi_2, t} \binom{t}{0} [I_d - I_s] \cdot f_1(t, d, \psi_1, \psi_2) \times$$
$$\times \sin \psi_1 \cdot \sin \psi_2 \, d\psi_1 \, d\psi_2,$$ (32)

where

$$P(|\tau| > t, \alpha) = \int_t^\infty f(\tau, \alpha)\, d\tau$$

is the probability that the length of the 'typical' edge of a random mosaic having direction α will be greater than t.

If a random mosaic is also isotropic then

$$\frac{\partial^2 p_0(t, \alpha)}{\partial \alpha^2} = 0$$

we arrive to a relation found in [6]. The main conclusion from (32) which can now be done in the general case is that the length distribution of a typical edge of a mosaic in direction α_0" can be found using the values of $p_0(t, \alpha)$ only in some neighborhood of α_0 and the distribution of $\{\mathscr{P}_I, \psi_i\}$ (see the introduction) on a test line of direction α_0.

References

1. Ambartzumian, R. V.: *Combinatorial Integral Geometry: with Applications to Mathematical Stereology*, Wiley, New York, 1982.
2. Ambartzumian, R. V.: 'Stochastic Geometry from the Standpoint of Integral Geometry', *Adv. Appl. Prob.* **9** (1977), 792–823.
3. Ambartzumian, R. V.: 'Probability Distributions in Stereology of Random Geometrical Processes', in *Recent Trends in Mathematics*, Reinhardbrunn (collection of papers), v. 50, BSB B.G. Teubner Verlagsgesellschaft, Leipzig, 1982, pp. 5–12.
4. Oganian, V. K.: 'Combinatorial Principles in Stochastic Geometry of Random Segment Processes', *Dokl. Akad. Nauk Arm. SSR* **68** (1979), 150–154.
5. Oganian, V. K.: 'On Palm Distributions of Processes of Lines in the Plane', Teubner Texte zur Mathematik, v. **65** 1984, pp. 124–132.
6. Oganian, V. K.: 'On a Distribution of the Length of the "Typical" Edge of a Random Tessellation', *Izv. Akad. Nauk Arm. SSR, ser. Math.* **19** (1984), 248–256.
7. Oganian, V. K.: 'Combinatorial Principles in Stochastic Geometry of Random Segment Processes', in [8], pp. 81–106.
8. Ambartzumian, R. V. (ed.): *Combinatorial Principles in Stochastic Geometry* (collection of papers), Publishing House of the Armenian Academy of Sciences, Yerevan, 1980.
9. Ambartzumian, R. V.: *Factorization in Integral and Stochastic Geometry*, Teubner Texte zur Mathematik, V. 65, 1984, pp. 14–33.
10. Santalo, L. A.: *Integral Geometry and Geometric Probability*, Addison-Wesley, 1976.
11. Kallenberg, O.: *Random Measures*, Akademie Verlag, Berlin, Reading, Mass., 1983.

Acta Applicandae Mathematicae **9** (1987), 83–95.
© 1987 by D. Reidel Publishing Company.

Randomizable Point Systems

G. S. SUKIASIAN
Institute of Mathematics, Armenian Academy of Sciences, pr. M. Bagramian 24ᵇ,
Yerevan 375019, U.S.S.R.

(Received: 24 June 1986)

Abstract. This paper gives a survey of the theory of point-set randomizations with respect to a group. It is shown that only the so-called k-lattices are randomizable with respect to the parallel translations and all motions of the R^n space. It is shown how the randomizability problems are connected with questions of the theory of discrete Lie groups and regular point systems. The results are applied to the construction of stationary random line processes.

AMS subject classifications (1980). 60D05, 60G55.

Key words. Stochastic geometry, random lattices, point processes, line processes, Palm distribution, regular point systems.

0. Introduction

The lattice ρ consisting of the points in the Euclidean space R^n with integer coordinates, is the simplest example of a point system randomizable with respect to the group T_n of parallel translations of the space R^n. We bear in mind the following property of the lattice ρ. We translate the lattice ρ by the random vector distributed uniformly within $[0, 1)^n$. We then obtain a random point process with distribution P which has the properties:

(a) P is invariant with respect to the group T_n,
(b) P is concentrated on the set $\{t\rho : t \in T_n\}$. ($t\rho$ is the image of ρ under the transformation t.)

To describe all point systems with such properties could be of great interest. It is natural to also consider similar problems for other groups. We give the corresponding definitions.

DEFINITION 0.1. A subset of the space R^n which has no concentration points is called a point system.

DEFINITION 0.2. Let G be a group of transformations of the space R^n. A point system ω is called G-randomizable if there exists a G-invariant random point process whose distribution is concentrated on the set of images $\{g\omega : g \in G\}$.

A further example of a randomizable point system is given by Siegel's theorem from integral geometry [8]. We denote by $\mathscr{A}_n$ the group of affine transformations of R^n which preserve the Lebesgue measure. According to Siegel's theorem, the lattice ρ consisting of the points of R^n with integer coordinates is $\mathscr{A}_n$-randomizable.

Work on randomizable point systems was started in the author's papers [10–12]. The present paper gives an up-to-date survey of the theory. We show how the randomizability problems are connected with questions of the theory of discrete Lie groups and regular point systems.

In the last section the results are applied to the construction of stationary random line processes on the plane.

Although the problems treated below are probabilistic in nature, it is more convenient to formulate some results in algebraic language. All the necessary concepts and results from the theory of Lie groups are given in Section 1.

1. Some Necessary Notions from the Theory of Discrete Lie Groups

1.1. We denote by U_n the group of all transformations of the space R^n into itself. All groups considered here are Lie subgroups of the universal group U_n. Let G be such a subgroup and ω be a point system. Consider the set $G(\omega)$ of transformations $g \in G$, for which $g\omega = \omega$. It is obvious that $G(\omega)$ is a subgroup of the group G for any $\omega \subset R^n$. Since ω has no concentration points, the subgroup $G(\omega)$ is necessarily discrete.

Later we will need the concept of the factor space G/H of the group G by the subgroup H: by definition, G/H is the set of the cosets gH, $g \in G$ with its factor topology (see [7]).

DEFINITION 1.1 ([7]). Discrete subgroup H of the group G is said to be of the lattice type if there exists a finite invariant measure on the factor space G/H.

PROPOSITION 1.1. *A point system ω is G-randomizable if and only if the discrete subgroup $G(\omega)$ of the group G is of the lattice type.*

To prove this it is sufficient to show that the factor space $G/G(\omega)$ and the set of the point systems $\{g\omega : g \in G\}$ are isomorphic. Let the coset $gG(\omega)$, $g \in G$ correspond to the point system $g\omega$.

It is obvious that this mapping is a homomorphism. Let us prove that it is injective. Let $g\omega = h\omega$; h, $g \in G$, let us show that $gG(\omega) = hG(\omega)$. We have $h^{-1}g \in G(\omega)$, hence $g \in hG(\omega)$. But $g \in gG(\omega)$ the cosets $gG(\omega)$ and $hG(\omega)$ have a common element g, hence they coincide. Vice-versa, from $gG(\omega) = hG(\omega)$ we obtain $h^{-1}g \in G(\omega)$ and $h^{-1}g\omega = \omega$, hence $h\omega = g\omega$. Proposition 1.1 is proved.

1.2. We have noted in the Introduction that the lattice ρ is T_n-randomizable. The factor space $T_n/T_n(\rho)$ is compact, it coincides with the n-dimensional fundamental parallelepiped of the lattice. Yet, as mentioned in [7], there exists a discrete lattice-type subgroup with a noncompact factor space. For a broad class of the so-called solvable Lie groups G [7], a discrete subgroup H is of the lattice type if and only if the factor space G/H is compact.

1.3. For any two subgroups G, H of the universal group U_n we consider the set

$$GH = \{gh : g \in G, h \in H\}.$$

GH can be a group itself. For example, let T_n be the group of parallel translations and ϕ_n be the group of rotations of the space R^n. Then $T_n \phi_n$ is the group of all Euclidean motions.

Consider such a subgroup G of the group U_n that $T_n G$ is a subgroup. If for each $tg \in T_n G$ there exists such $t' \in T_n$ that

$$tg = gt', \tag{1.1}$$

then the group $T_n G$ is called semicommutative.

1.4. As mentioned in [7], all semicommutative Lie groups are solvable. Using Subsection 1.2 and Proposition 1.1, we obtain the following proposition.

PROPOSITION 1.2. *Let $T_n G$ be a semicommutative Lie group. The point system ω is $T_n G$-randomizable if and only if the factor space $T_n G / T_n G(\omega)$ is compact.*

COROLLARY. *For any compact group G all point systems are G-randomizable.*

1.5. Later on we shall need the following result from the theory of Lie groups [7]. Let G be a Lie group and H_1, H_2 its closed subgroups, $H_1 \subset H_2$. If H_1 is of the lattice type, then H_2 is of lattice type too.

2. General Properties

In this section we discuss the question of how new randomizable point systems can be obtained from a given randomizable system.

2.1. Let ω be a G-randomizable point system. The following proposition states a sufficient condition for G-randomizability of $h\omega$ for a $h \in U_n$.

PROPOSITION 2.1. *Let the transformation $h \in U_n$ have the property $hG = Gh$. A point system ω is G-randomizable if and only if $h\omega$ is G-randomizable.*

Proof. All subgroups isomorphic to a lattice-type subgroup are themselves of the lattice type. Using Proposition 1.1, it is enough to show that the discrete subgroups $G(h\omega)$ and $G(\omega)$ are isomorphic for any point system ω. To each transformation $g \in G$ corresponds a $f(g) \in G$ for which $hg = f(g)h$. The mapping f is a homomorphism, since for any g_1, $g_2 \in G$ we have

$$f(g_1 g_2)h = hg_1 g_2 = f(g_1)hg_2 = f(g_1)f(g_2)h.$$

If $g \in G(\omega)$ then $f(g)h\omega = hg\omega = h\omega$. Hence, $f(g) \in G(h\omega)$. In the same way we obtain that if $g \in G(h\omega)$, then $f^{-1}(g) \in G(\omega)$. Hence, $G(\omega)$ and $G(h\omega)$ are isomorphic subgroups.

2.2. COROLLARY. *Let the point system ω be G-randomizable. For any $g \in G$ the point system $g\omega$ is also G-randomizable.*

2.3. DEFINITION 2.1. The point system ω is called a lattice if ω is obtained by affine transformation from the set ρ of the points of R^n with integer coordinates.

We have noted in the Introduction, that the lattice ρ is T_n-randomizable. For any affine transformation a we have $aT_n = T_n a$. It follows from Proposition 2.1 that all lattices in the space R^n are T_n-randomizable.

2.4. When is the union of randomizable point systems also randomizable?

PROPOSITION 2.2. *Let ω be a G-randomizable point system and $g \in G$ be a transformation with the property $gG(\omega) = G(\omega)g$. Then the point system $\omega \cup g\omega$ is G-randomizable too.*

Proof. According to Subsection 1.5, it is sufficient to show that $G(\omega) \subset G(\omega \cup g\omega)$. For any $h \in G(\omega)$ exists such $h' \in G(\omega)$ that $hg = gh'$. We have

$$hg\omega = gh'\omega = g\omega.$$

Hence,

$$h(\omega \cup g\omega) = h\omega \cup hg\omega = \omega \cup g\omega, \quad h \in G(\omega \cup g\omega).$$

2.5. DEFINITION 2.2. Let ω be a lattice and $t_1, \ldots, t_k$ be some parallel translations. The union $\bigcup_{i=1}^{k} t_i\omega$ is called a k-lattice.

It follows from Proposition 2.2, the commutativity of the group T_n and T_n-randomizability of all lattices (see Subsection 2.3), that all k-lattices in the space R^n are T_n-randomizable. Later we will show that the class of T_n-randomizable point systems is confined to k-lattices.

2.6. Let us show, that each randomizable point system can be represented as a finite union of so-called regular point systems. Below, Gx, $x \in R^n$ denotes a point set $\{gx \in R^n : g \in G\}$. From the definition of a discrete group $G(\omega)$ (see Subsection 1.1) it follows that for any $x \in \omega$

$$G(\omega)x \subset \omega. \tag{2.1}$$

Since each point system ω contains no more than a denumerable set of points, we have $\omega = \bigcup_i G(\omega)x_i$ where $\{x_i\}$ is a finite or denumerable sequence of points from ω.

PROPOSITION 2.3. *Let $T_n G$ be a semicommutative Lie group. For any $T_n G$-randomizable point system ω, there exists such a finite sequence $x_1, \ldots, x_k \in \omega$ that $\omega = \bigcup_{i=1}^{k} T_n G(\omega)x_i$.*

Proof. Let us suppose that

$$\omega = \bigcup_{i=1}^{\infty} T_n G(\omega)x_i, \quad x_j \notin T_n G(\omega)x_i, j > i.$$

Let t_i be the parallel translation for which $t_i O = x_i$, where O is the origin. We have

$$T_n G(\omega) t_i O = T_n G(\omega) x_i \neq T_n G(\omega) x_j = T_n G(\omega) t_j O.$$

Hence,

$$T_n G(\omega) t_i \neq T_n G(\omega) t_j, \quad i \neq j.$$

According to Proposition 1.2 the factor space $T_n G / T_n G(\omega)$ is compact. Hence, there exists a convergent sequence of the cosets: $T_n G(\omega) t_{i_k} \to T_n G(\omega) t_0 g_0$. The sequence $x_{i_k} = t_{i_k} O \in \omega$ has a concentration point at $t_0 g_0 O$. This contradicts the condition that ω is a point system.

2.7. COUNTEREXAMPLE. Let us show, that Proposition 2.3 does not hold for an arbitrary group (which has not the form $T_n G$). Consider the group ϕ_n of rotations of R^n around O. The group ϕ_n is compact and, according to Subsection 1.4, all point systems in R^n are ϕ_n-randomizable. Let ω be the system of points with positive integer coordinates. The discrete subgroup $\phi_n(\omega)$ is trivial, for it contains only identity transformation. For any $x \in \omega$ we have $\phi_n(\omega) x = x$ and ω cannot be represented as a finite union $\cup_i \phi_n(\omega) x_i$.

2.8. Consider the point system $\tau = G(\omega) O$ where G is some subgroup of the group U_n and ω is a point system, $O \in \omega$.

Since $G(\omega)$ is a subgroup we have $g G(\omega) = G(\omega)$ for each $g \in G(\omega)$. Hence, $g G(\omega) O = G(\omega) O$ that is $g \in G(\tau)$. Thus, for any ω and G we have

$$G(\omega) \subset G(G(\omega) 0). \tag{2.2}$$

It follows from Proposition 1.1, the inclusion (2.2) and Subsection 1.5, that for any G-randomizable ω, the point system $\tau = G(\omega) O$ is G-randomizable too.

2.9. EXAMPLE. Let us give the example of a point system, for which the inclusions (2.1) and (2.2) are strict. Consider the 2-lattice $\omega = \rho \cup t\rho$ in the space R^n, where ρ is the lattice of points with integer coordinates, $t \in T_n$ is some parallel vector translation with irrational coordinates. Consider the group $T_n \phi_n$ of Euclidean motions of the space R^n. The discrete subgroup $T_n \phi_n(\omega)$ coincides with the group $T_n(\rho)$ of translations (no rotations!) by vectors with integer coordinates. We have $T_n \phi_n(\omega) O = T_n(\rho) O = \rho \subset \omega$ (compare with (2.1)), obviously this inclusion is strict. We have also

$$T_n \phi_n(T_n \phi_n(\omega) O) = T_n \phi_n(\rho) = T_n(\rho) \phi_n(\rho) \supset T_n(\rho) = T_n \phi_n(\omega)$$

(compare with (2.1)). This inclusion is also strict because the subgroup $\phi_n(\rho)$ is nontrivial. In particular, $\phi_2(\rho)$ is the discrete group of the rotations by angles divisible by $\pi/2$.

3. Regular Point Systems

DEFINITION 3.1 (see [3]). A point system ω, $O \in \omega$ is called regular with respect to a group G, if $G(\omega) O = \omega$.

3.1. Consider the point system $\tau = G(\omega)O$. Using (2.2), we obtain $G(\tau)O \supset G(\omega)O = \tau$. According to (2.1), we have $C(\tau)O \subset \tau$. Hence, $G(\tau)O = \tau$. So for any $\omega \subset R^n$ the point system $G(\omega)O$ is regular with respect to each group G.

3.2. PROPOSITION 3.1 ([3]). *A point system $\omega \subset R^n$ is regular with respect to the group T_n of parallel translations of the space R^n if and only if ω is a lattice.*

THEOREM 1. *A point system $\omega \subset R^n$ is T_n-randomizable if and only if ω is a k-lattice.*

Proof. The sufficiency follows from Subsection 2.5. According to Proposition 2.3, we have $\omega = \bigcup_{i=1}^{k} T_n(\omega)x_i$. From Subsection 3.1 and Proposition 3.1, it follows that $T_n(\omega)O$ is a lattice. But $T_n(\omega)x_i$ is obtained from the lattice $T_n(\omega)O$ by means of translation by the vector x_i, $i = 1, 2, \ldots, k$. Hence, ω is a k-lattice.

3.3. We now consider the group $T_n\phi_n$ of Euclidean motions of the space R^n. Note that the group $T_n\phi_n$ is semicommutative.

PROPOSITION 3.2 ([3]). *All point systems regular with respect to the Euclidean group $T_n\phi_n$ (the so-called crystallographic sets) are k-lattices (the converse is not true).*

THEOREM 2. *A point system $\omega \subset R^n$ is $T_n\phi_n$-randomizable if and only if ω is a k-lattice.*

Proof. Consider a stationary (T_n-invariant) random point process whose distribution is concentrated on the T_n-images of some k-lattice (such process exists according to Theorem 1). This process can be made isotropic using a random rotation distributed uniformly within the compact group ϕ_n. Hence, it follows that all k-lattices are $T_n\phi_n$-randomizable.

The necessity is proved similarly to Theorem 1 using Propositions 2.3 and 3.2.

3.4. So using Subsections 2.8, 3.1, and Proposition 2.3, we obtain the following algorithm for constructing randomizable point systems. First we find all the regular point systems for the given group G. Then we select such regular point systems which are G-randomizable. Their finite sums make up the set of all G-randomizable point systems.

Below we construct a subgroup T_2C of affine transformations for which there exist regular (and randomizable) point systems which are not k-lattices.

3.5. Note that if ω is a G_1-randomizable point system, then it does not follow that ω is G_2-randomizable, even if we assume that $G_1 \subset G_2$ or $G_2 \subset G_1$. So Theorems 1 and 2 are independent.

4. Discrete Palm Distributions

The proofs of all the propositions of this section can be found in [10].

4.1. In the space R^n we consider a stationary (T_n-invariant) random point process of finite intensity. By π we denote its Palm distribution (see [5]).

Roughly speaking, π is the conditional distribution of the point process, given that a point of its realization coincides with O.

PROPOSITION 4.1. (a) *For any realization ω containing the points O and x we have*

$$\pi(\{\omega\}) = \pi(\{t_x(\omega)\}),$$

where $t_x \in T_n$ is translation by vector $x \in R^n$.

(b) *If $\pi(\{\omega\}) > 0$ then the realization ω is a k-lattice and $k \leqslant \pi(\{\omega\})^{-1}$.*

DEFINITION 4.1. Let ω be a G-randomizable point system. A G-invariant random point process concentrated on the set of images $\{g\omega : g \in G\}$ is called a G-invariant randomization of the point system ω.

4.2. Consider a k-lattice $\omega = \bigcup_{i=1}^{k} t_{x_i}\rho$ where ρ is a lattice. Let $O \in \rho$, hence $x_i \in \omega$, $i = 1, 2, \ldots, k$. The Palm distribution π of the T_n-invariant randomization of k-lattice ω is concentrated on the set $\{t_{x_i}^{-1}\omega, i = 1, \ldots, k\}$ and, according to Proposition 4.1(a),

$$\pi(\{t_{x_i}^{-1}\omega\}) = \frac{1}{k}, \quad i = 1, 2, \ldots, k.$$

Hence, the finite and denumerable mixtures of T_n-invariant randomizations of k-lattices have discrete Palm distributions. The distribution of a point process we call discrete, if it is concentrated on finite or countably many realizations. Using Proposition 4.1(b) we obtain the following theorem.

THEOREM 3. *The Palm distribution of a stationary random point process P in the space R^n is discrete if and only if the process P is a finite or denumerable mixture of T_n-invariant randomizations of k-lattices.*

4.3. Similar results have also been obtained for the group $T_n\phi_n$ of all Euclidean motions. In this case, some realization 'bundles' $\Delta(\omega) = \{\varphi\omega : \varphi \in \phi_n\}$ have positive probabilities. Now let π be Palm distribution of a homogeneous and isotropic (i.e., $T_n\phi_n$-invariant) random point process with finite intensity. If $\pi(\Delta(\omega)) > 0$, then the realization ω is a k-lattice.

THEOREM 4. *A homogeneous and isotropic random point process is a finite (denumerable) mixture of the $T_n\phi_n$-invariant randomizations of k-lattices if and only if its Palm distribution is concentrated on a finite (denumerable) union of realization bundles.*

5. Group $T_2\mathbb{C}$

5.1. Let us denote by $\mathbb{C}$ the subgroup of affine transformations C of the plane with matrix

$$\begin{pmatrix} 1 & c \\ 0 & 1 \end{pmatrix}, \quad c \in R^1.$$

Then T_2C is the group of the following affine transformations tC:

$$\left. \begin{aligned} x' &= x + cy + a \\ y' &= y + b \end{aligned} \right\} a, b, c \in R^1. \tag{5.1}$$

According to Equations (5.1) we have

$$tC = Ct', \quad \text{where } t' = (a - cb; b).$$

So the group T_2C satisfies the condition (1.1) and, hence, T_2C is a solvable group (but T_2C is not Abelean).

Note that the group T_2C participates in the construction of the so-called Galilei geometry [13].

5.2. Consider the plane lattice ρ of points with integer coordinates. We transform the lattice ρ by (5.1) with random independent parameters a, b, c, which are distributed uniformly within $[0, 1]$. We obtain a T_2C-invariant randomization of the lattice ρ. Hence, the lattice ρ is a T_2C-randomizable point system (see [12]).

5.3. Let h be an affine transformation of the plane by the matrix $\begin{pmatrix} h_x & 0 \\ 0 & h_y \end{pmatrix}$ (axial compression). For any h_x, $h_y \in R^1$ we have $hT_2C = T_2C h$. According to Proposition 2.1, the lattice $h\rho$ is a T_2C-randomizable point system.

5.4. DEFINITION 5.1. A plane lattice ω is called a lattice of type I if ω contains the points with the same y-coordinates. Otherwise, ω is called a lattice of type II.

The transformations from the group T_2C and axial compressions do not change the type of lattices, but the rotations can do. Any type-I lattice will be obtained from the lattice ρ using axial compression and transformation $tC \in T_2C$. It follows from Subsections 5.3, 2.2, and Proposition 2.1.

PROPOSITION 5.1. *All type-I lattices on the plane are T_2C-randomizable point systems.*

Note that the type-II lattices are not T_2C-randomizable.

5.5. Each point system regular with respect to a group G_1 is also regular with respect to any broader group G_2, $G_1 \subset G_2$. Hence, the class $\mathscr{F}$ of point systems regular with respect to the group T_2C contains all lattices (compare with Proposition 3.1). The complete description of $\mathscr{F}$ is given in Theorem 6.

5.6. In the space R^n there exist several analogues of the group T_2C. For example, in the space R^3 there exist two such groups: the group $T_3C_3^1$ of the following affine transformations:

$$x' = x + cz + a, \qquad y' = y + b, \qquad z' = z + d,$$

where $a, b, c, d \in R^1$, and the group $T_3 C_3^2$ of affine transformations of the form

$$x' = x + c_1 z + a, \qquad y' = y + c_2 z + b, \qquad z' = z + d,$$

where $a, b, c_1, c_2, d \in R^1$.

Now let ρ be the lattice consisting of the points in the space R^3 with integer coordinates. All lattices obtained from ρ by axial compressions and by transformations from $T_3 C_3^i$ are $T_3 C_3^i$-randomizable point systems, $i = 1, 2$.

6. Parabolic Inversion

6.1. The symmetries with respect to the parabolas (the so-called inversions) are basic in Galilei's geometry (see [13]).

DEFINITION 6.1. The nonaffine transformation of the plane

$$\left.\begin{aligned} x' &= ry^2 - x \\ y' &= y \end{aligned}\right\} r \in R^1 \tag{6.1}$$

is called parabolic inversion with parameter r and is denoted by $\mathcal{T}_r$.

6.2. Let Z be a parabola with curvature r_0 and with axis parallel to the X-axis (i.e., Z is described by the equation $x = r_0/2 (y - y_0)^2 + x_0$). Then $\mathcal{T}_r Z$ is a parabola with curvature $2r - r_0$.

The parabolas with the same curvature and axis are called concentric parabolas. The $\mathcal{T}_r$-images of two parallel lines (the parabolas with zero-curvature) are concentric parabolas both having curvature $2r$ (except the lines parallel to the X-axis).

6.3. Using Equations (5.1) and (6.1), we obtain the following proposition.

PROPOSITION 6.1. *For any* $r \in R^1$ *we have*

$$\mathcal{T}_r = \mathcal{T}_r^{-1} \tag{6.2}$$
$$\mathcal{T}_r T_2 \mathbf{C} = T_2 \mathbf{C} \mathcal{T}_r \tag{6.3}$$

According to (6.3) and Proposition 2.2, we have

PROPOSITION 6.2. *A point system* ω *is* $T_2\mathbf{C}$*-randomizable if and only if its inversion* $\mathcal{T}_r \omega$ *is* $T_2\mathbf{C}$*-randomizable for any* $r \in R^1$.

6.4. According to (6.3) for any transformation $t\mathbf{C} \in T_2\mathbf{C}$ there exists such $t'\mathbf{C}' \in T_2\mathbf{C}$ that $\mathcal{T}_r t\mathbf{C} = t'\mathbf{C}'\mathcal{T}_r$. If $t\mathbf{C} \in T_2\mathbf{C}(\omega)$ then

$$t'\mathbf{C}'\mathcal{T}_r \omega = \mathcal{T}_r t\mathbf{C}\omega = \mathcal{T}_r \omega$$

i.e., $t'\mathbf{C}' \in T_2\mathbf{C}(\mathcal{T}_r \omega)$. Hence

$$\mathcal{T}_r[T_2\mathbf{C}(\omega)] = [T_2\mathbf{C}(\mathcal{T}_r \omega)]\mathcal{T}_r. \tag{6.4}$$

Let ω be a point system regular with respect to the group T_2C, $O \in \omega$. According to (6.4) and since $\mathcal{T}_e O = 0$, we have

$$[T_2C(\mathcal{T}_r\omega)]O = \mathcal{T}_r[T_2C(\omega)]O = \mathcal{T}_r\omega$$

hence, $\mathcal{T}_r\omega$ is also regular. Using (6.2) we obtain the following proposition.

PROPOSITION 6.3. *A point system ω is regular with respect to the group T_2C if and only if its inversion $\mathcal{T}_r\omega$ for any $r \in R^1$ is regular with respect to the group T_2C.*

6.5. In the space R^n there exist inversions for which propositions similar to 6.2 and 6.3 hold. For example, in the space R^3 the group $T_3C_3^1$ corresponds to the following inversion:

$$x' = rz^2 - x, \qquad y' = y, \qquad z' = z. \quad r \in R^1$$

and the group $T_3C_3^2$ corresponds to the inversion

$$x' = r_1z^2 - x, \qquad y' = r_2z^2 - y, \qquad z' = z, \quad r_1, r_2 \in R^1$$

7. Parabolic Lattices

7.1. DEFINITION 7.1. Let ω be a planar lattice and $\mathcal{T}_r$ be a parabolic inversion with parameter $r \in R^1$. The point system $\mathcal{T}_r\omega$ is called a parabolic lattice. If ω is type j lattice (see Definition 5.1) then $\mathcal{T}_r\omega$ is called type j parabolic lattice, $j = 1, 2$.

Using Propositions 5.1 and 6.2 we obtain Theorem 5.

THEOREM 5. *All type-I parabolic lattices are T_2C-randomizable point systems.*

7.2. We have noted in Subsection 5.5 that all planar lattices are regular with respect to the group T_2C. According to Proposition 6.3, all parabolic lattices are also regular. In [12] we have shown that the inverse proposition is also correct.

THEOREM 6. *A point system ω is regular with respect to the group T_2C if and only if ω is a parabolic lattice (of any of the two types).*

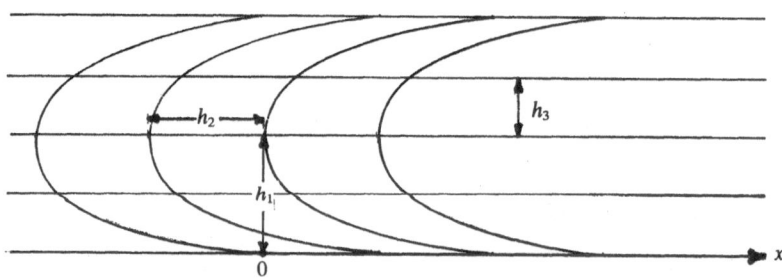

Fig. 1.

7.3. A parabolic inversion does not change the lines parallel to X-axis. The distances between the points lying on such lines are preserved. Using Subsection 6.2, we obtain the following proposition.

PROPOSITION 7.1. *A type-I parabolic lattice is a set of intersection points of equidistant parallel lines and concentric parabolas, whose apices form a one-dimensional lattice (see Figure 1).*

8. Davidson's Problem

8.1. The above results are applied to the solution of a problem due to Davidson [2] which aroused considerable interest [1,4,6,9,11]. Does there exist a stationary (i.e., T_2-invariant) second-order random line process on the plane which is not a Cox (i.e., double stochastic Poisson) process and whose realizations do not contain the parallel lines with probability 1? This problem was solved by Kallenberg in [4] where an example of such process was constructed. We construct a whole family of such examples which contains the Kallenberg's example as a particular case.

8.2. Let us describe the line processes as point processes. A line g on the plane is defined by coordinates (x, y) where x is the abscissa of intersection point of line g with the X-axis, $y = \cot \varphi$, φ is the angle between g and X-axis (Figure 2). If P_g is a random line process whose realizations do not contain lines parallel to the X-axis with probability 1, then P_g corresponds to a random point process P on the parametric plane.

The parallel translation of the plane by vector (a, c) corresponds to the following transformation of the parametric plane, Figure 2.

$$x' = x + cy + a, \qquad y' = y. \tag{8.1}$$

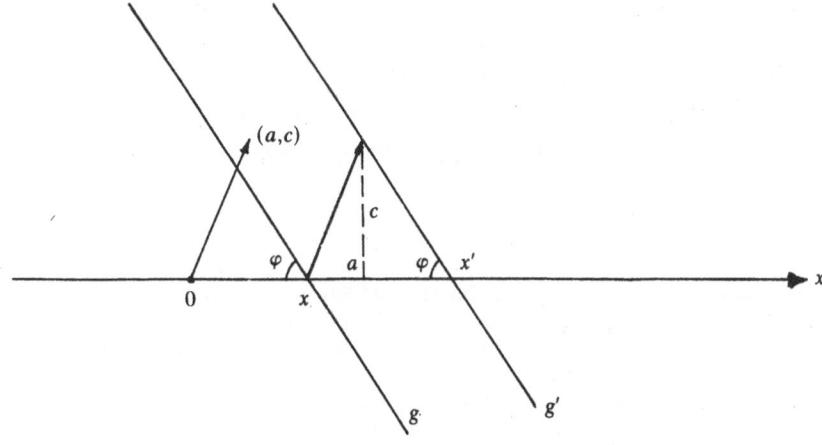

Fig. 2.

The line process P_g happens to be stationary if and only if the corresponding point process P is invariant with respect to the group B of affine transformations (8.1). It is obvious that the group B is a subgroup of the group T_2C (see Section 5).

Note that parallel lines correspond to the points on the parametric plane with the same y-coordinates. Hence, to solve Davidson's problem it is enough to construct a non-Coxian T_2C-invariant random point process, on the parametric plane, whose realizations do not contain the points with the same Y-coordinates with probability 1. The latter condition does not permit us to use the T_2C-invariant point process concentrated on the type-I parabolic lattices (this process exists according to Theorem 5).

8.3. The measure μ_r on R^2 is called an inversion of the measure μ on the plane if $\mu_r(A) = \mu(\mathcal{T}_r A)$ for any Borel set $A \subset R^2$. Similarly, the inversion P_r of a point process P is defined by the equation

$$P_r(A) = P(\{\omega : \mathcal{T}_r \omega \in A\}).$$

According to Proposition 6.1, we have Proposition 8.1.

PROPOSITION 8.1. *A point process P on the plane is T_2C-invariant if and only if its inversion P_r is T_2C-invariant for any $r \in R^1$.*

From the properties of the parabolic inversion we obtain Proposition 8.2.

PROPOSITION 8.2. *Let P be a Cox process with a governing measure μ. For any $r \in R^1$ the inversion P_r is a Cox process with the governing measure μ_r.*

8.4. In [4] Kallenberg constructed his example P' using Siegel's theorem (see Introduction). The line process P' satisfies Davidson's conditions and the corresponding point process P is concentrated on the pieces of type-II lattices. The inversion P_r of point process P will be concentrated on the pieces of type-II parabolic lattices. Hence, the realizations of the corresponding line process P'_r will not contain parallel lines with probability I.

According to Propositions 8.1 and 8.2, for any $r \in R^1$ the point process P_r is not Coxian but P_r is T_2C-invariant. Hence, the line process P'_r is stationary and we conclude that the family $\{P'_r\}$, $r \in R'$ provides further examples for Davidson's problem. For $r = 0$ we obtain the example of Kallenberg.

References

1. Ambartzumian, R. V.: 'Factorization in Integral and Stochastic Geometry', *Stochastic Geometry, Geometric Statistics, Stereology*, Teubner-Texte z. Math. v. 65, 1984.
2. Davidson, R.: Construction of Line Processes, Second Order Properties', *Izw. AN Armenian SSR, ser. Math.* **5** (1970), 219–234.
3. Hilbert, D. and Cohn-Vossen, S.: *Anschauliche Geometrie*, Berlin, 1932.
4. Kallenberg, O.: 'A Counterexample to R. Davidson's Conjecture on Line Processes', *Math. Proc. Camb. Phil. Soc.* **82** (1977), 301–307.

5. Matthes, K., Kerstan, J., and Mecke, J.: *Infinitely Divisible Point Processes*, Wiley, Chichester, 1978.
6. Mecke, J.: An Explicit Description of Kallenberg's Lattice Type Point Process', *Math. Nachr.* **89** (1979), 185–195.
7. Raghunathan, M. S.: *Discrete Subgroups of Lie Groups*, Springer-Verlag, Berlin, Heidelberg, New York, 1972.
8. Santalo, L. A.: *Integral Geometry and Geometric Probability*, Addison-Wesley, Reading, Mass., 1976.
9. Harding, E. F., and Kendall, D. G. (eds.), *Stochastic Geometry*, Wiley, New York, 1971.
10. Sukiasian, G. S.: 'On Characterization of Random Lattices', *Soviet J. Contemporary Math. Anal.* **20** (1985).
11. Sukiasian, G. S.: 'Invariant Randomizations of Lattices and the Davidson Problem' (Russian), *Abstracts of Communications IV Internat. Vilnius Conf. of Probab. Theory and Math. Stat.* v.3, 1985, pp. 162–163.
12. Sukiasian, G. S.: 'Randomizable Point Systems and Parabolic Lattices' (to appear).
13. Yaglom, I. M.: *Galilei's Relativity Principle and Non-Euclidean Geometry* (Russian), Nauka, Moscow, 1969.

Acta Applicandae Mathematicae **9** (1987) 97–102.
© 1987 *by D. Reidel Publishing Company.*

Expected Convex Hulls, Order Statistics, and Banach Space Probabilities[*]

RICHARD A. VITALE
Department of Mathematics, Claremont Graduate School, Claremont, CA 91711, U.S.A.

(Received: 22 April 1986)

Abstract. Using the expectation of a particular random set, we present a multivariate extension of a characterization theorem involving extreme order statistics.

AMS subject classifications (1980). Primary: 60D05; secondary: 52A22, 60B11, 62G30.

Key words. Abstract Wiener space, convex hull, extreme order statistic, order statistics, random sets, Wiener measure.

1. Introduction

The characterization of probability distributions by moments of associated order statistics apparently was first studied by Hoeffding (1953), who was interested in applications to rank order tests for hypotheses of randomness (Hoeffding, 1951; Terry, 1952). Since then, there has been considerable activity in extending and refining results in this area. As shown in a survey by Galambos and Kotz (1978), one natural line of inquiry, however, seems to have been elusive, namely multivariate extensions. Here we present a result in this direction by looking at the problem in a slightly different way and using ideas from the theory of random sets. We emphasize that it is the latter machinery and in particular the notion of the expectation of a random set which provides a natural (and immediate) extension. This expectation appeared in a law of large numbers for random sets (Artstein and Vitale, 1975), which has subsequently been extended (Artstein and Hansen, 1985), and supplemented with a central limit theorem (Weil, 1982). The centroid known as the Steiner point can also be understood as an expectation (e.g., Vitale, 1985).

The following univariate result is our point of departure. It can ultimately be traced to Hoeffding (1953) and has been rediscovered several times (for discussion and related results, see Galambos and Kotz, 1978, pp. 53–63). The reader may like to provide a proof in passing.

THEOREM 1. *Let* $X_1, X_2, \ldots$ *be an iid sequence of random variables with* $E|X_1| < \infty$. *Then the common distribution of the* X_i's *is determined by the increasing sequence* $\{E \max\{X_1, \ldots, X_n\}\}_{n=1}^{\infty}$.

[*] Supported in part by National Science Foundation grant DMS 8603944.

Our aim is a multivariate extension. In the next section, we review briefly some of the structure of random sets. The extension is then given in Section 3. There is no added difficulty in taking a separable Banach space as the setting. Indeed, this allows us to draw some interesting and unexpected connections with previous work. In the case of Gaussian measures, there are connections with the Wiener process (Cameron and Martin, 1944), abstract Wiener spaces (Segal, 1956, 1958; Gross, 1965) and laws of the iterated logarithm (Strassen, 1974).

2. Random Sets

Let $\mathbf{B}$ be a separable Banach space, and let $\mathbf{K}$ denote its collection of nonempty compact, convex subsets. We use conv($\cdot$) to denote convex hull. $\mathbf{K}$ can be topologized by the Hausdorff metric, given by

$$h(K_1, K_2) = \inf\{\epsilon > 0 \,|\, K_1 \subseteq K_2 + \epsilon B, \ K_2 \subseteq K_1 + \epsilon B\}$$

and becomes a complete, separable metric space. Here '+' denotes Minkowski summation of sets, i.e., $K + L = \{x + y \,|\, x \in K, y \in L\}$, scalar multiplication is defined as usual, and B is the closed unit ball of $\mathbf{B}$. A random compact, convex set, or simply random set, $\mathbf{X}$ is a map from a probability space into $\mathbf{K}$ which is measurable in the Borel sense.

Analysis of random sets is aided by an attractive embedding. Let S be the unit sphere in the dual space $\mathbf{B}^*$. To each set $K \in \mathbf{K}$, we associate its support function $s_K \colon S \to \mathbf{R}$, given by

$$s_K(X^*) = \sup\{x^*(x) \,|\, x \in K\}.$$

This identification is unique and permits $\mathbf{K}$ to be isometrically embedded in $C(S)$, the space of (norm) continuous functions of S. We record some features for later use:

$$K = L \Leftrightarrow s_K = s_L, \qquad K \subseteq L \Leftrightarrow s_K \leqslant s_L,$$

$$s_{K+L} = s_K + s_L, \qquad s_{\alpha K} = \alpha s_K, \quad \alpha \geqslant 0,$$

$$\|K\| \equiv \max\{\|x\| \,|\, x \in K\} = \|s_K\| \quad \text{(uniform norm)},$$

$$h(K, L) = \|s_K - s_L\|.$$

Given a random set $\mathbf{X}$, we may evidently regard it as a stochastic element of $C(S)$ via its support function $s_\mathbf{X}$.

We turn next to a discussion of expectations. Within $C(S)$, a random element possesses a Bochner expectation iff $E\|s_\mathbf{X}\| = E\|\mathbf{X}\| < \infty$. Since $|s_\mathbf{X}(x^*)| \leqslant \|s_\mathbf{X}\|$ for all $x^* \in S$, this condition assures the existence of $E s_\mathbf{X}(x^*)$ for all $x^* \in S$. Further, it is true that $E s_\mathbf{X}(x^*)$, as a function of x^*, is the support function of some compact, convex set. We define this set to the expectation of $\mathbf{X}$, written $E\mathbf{X}$, and which evidently satisfies

$$s_{EX}(x^*) = E s_X(x^*) \quad \forall x^* \in S.$$

Further details can be found in Artstein and Vitale (1975) and Giné and Hahn (1985). We mention that expectations of random sets, even nonconvex ones, can be formulated in terms of the various integrals of set-valued functions (see, for instance, Aumann, 1965; Artstein, 1974; Debreu, 1967; Artstein and Burns, 1974). The support function approach is the best choice for our purposes here.

3. Multivariate Extension

In extending Theorem 1 to more than one dimension, we first indicate our method by replacing the extreme order statistic in a one dimension sample with the convex hull of the sample, $\text{conv}\{X_1, \ldots, X_n\} = [\min\{X_1, \ldots, X_n\}, \max\{X_1, \ldots, X_n\}]$. The expectation (when it exists) of this random interval is the interval $[E \min\{X_1, \ldots, X_n\}, E \max\{X_1, \ldots, X_n\}]$. By Theorem 1, these intervals determine the underlying distribution. In higher dimensions, expectations of convex hulls, or *moment bodies* as we shall call them, continue to carry decisive distributional information. Convex hulls have been viewed before as random sets (for the purpose of various asymptotic results) in Fisher (1971), Eddy (1980), and Eddy and Gale (1981).

We now state the generalization of Theorem 1.

THEOREM 2. *Let* $X_1, X_2, \ldots$ *be an iid sequence of random elements in a separable Banach space* **B** *with Bochner expectation. Then the common distribution of the* X_i's *is determined by the nested sequence of compact, convex moment bodies* $\{E \text{ conv}\{X_1, \ldots, X_n\}\}_{n=1}^{\infty}$.

Proof. Recall that $X_1, X_2, \ldots, X_n, \ldots$ are iid elements of a separable Banach space **B**. The convex hull of the first n, $\mathbf{X}_n = \text{conv}\{X_1, \ldots, X_n\}$, satisfies the measurability criterion for being a random set. Since

$$\|\mathbf{X}_n\| = \max\{\|X_1\|, \ldots, \|X_n\|\} \leq \Sigma \|X_i\|,$$

we see easily that

$$E\|X_1\| < \infty \Rightarrow E\|\mathbf{X}_n\| \leq n E\|\mathbf{X}_1\| < \infty.$$

Consequently, for each n, we have the existence of the moment body $E\mathbf{X}_n = E \text{ conv}\{X_1, \ldots, X_n\}$ as an element of **K**.

Nesting of moment bodies follows from the sequence of implications

$$\mathbf{X}_n \subseteq \mathbf{X}_{n+1} \Rightarrow s_{\mathbf{X}_n} \leq s_{\mathbf{X}_{n+1}} \Rightarrow E s_{\mathbf{X}_n} \leq E s_{\mathbf{X}_{n+1}} \Rightarrow s_{E\mathbf{X}_n} \leq s_{E\mathbf{X}_{n+1}} \Rightarrow E\mathbf{X}_n \subseteq E\mathbf{X}_{n+1}.$$

It remains to show that the common distribution of the X_i's is determined by the moment body sequence or, equivalently, by the sequence of their support functions $\{E s_{\mathbf{X}_n}\}_{n=1}^{\infty}$. Recall that $E s_{\mathbf{X}_n}$ is the Bochner expectation of $s_{\mathbf{X}_n}$ in $C(S)$

and upon evaluation at x^* yields $s_{EXn}(x^*)$. Here we reverse the emphasis. Fix x^* and let n vary;

$$\{s_{EX_n}(x^*)\}_{n=1}^\infty = \{E \max\{x^*(X_1), \ldots, x^*(X_n)\}\}_{n=1}^\infty.$$

By Theorem 1, these values determine the distribution of $x^*(X_1)$. This is true for all $x^* \in S$ so that the probability content of each half-space and, hence, the underlying probability measure is determined.

4. Remarks

As promised, once the appropriate machinery is in place the statement and proof of the extension are direct. We surmise that this signals a natural viewpoint with a wider utility, especially in problems relying on the simultaneous behavior of linear functionals. For now, we turn to some other comments, the first of which is that, in a special case of Theorem 2, we have geometrically characterized a well-known object.

(1) In the case of a Gaussian measure on a Banach space, moment bodies take a particularly nice form. Recall that $x \in E \operatorname{conv}\{X_1, \ldots, X_n\}$ if and only if

$$x^*(x) \le E \max\{x^*(X_1), \ldots, x^*(X_n)\} \forall x^*.$$

With mean zero this occurs if and only if

$$x^*(x) \le \gamma_n \sqrt{\operatorname{Var} x^*(X_1)} \forall x^*$$

where we have set $\gamma_n = E \max\{Z_1, \ldots, Z_n\}$ for independent $N(0, 1)$ arguments. Setting

$$K = \{x \mid x^*(x) \le \sqrt{\operatorname{Var} x^*(X_1)} \forall x^*\}$$

yields $\gamma_n K$ as the nth moment body.

(*Question*: beyond the Gaussian case, when do moment bodies scale in this way?)

The set K has appeared earlier in the literature in connection with different problems. Apparently its first appearance is in the seminal paper of Cameron and Martin (1944) on the Wiener process, where

$$K = \left\{x \in C[0, 1] \mid x(t) = \int_0^t y(s)\,ds, \int_0^1 y^2(s)\,ds \le 1\right\}.$$

Later, in the theory of abstract Wiener spaces (Segal, 1956, 1958; Gross, 1965), K figures as the unit ball of a Hilbert space associated with an underlying measure. In a parallel development, it has also appeared in laws of the iterated logarithm (Strassen, 1964).

Novel aspects of the present study are that the compactness of K follows from its role as an expected set and that the general theory proceeds under the assumption of a first moment only. It may be that this holds potential for extending earlier results.

Some further particular cases follow.

(a) If $\mathbf{B} = R^d$ and the distribution is $N(0, \Sigma)$, then $K = \{\Sigma p \,|\, p \in R^d, p^T \Sigma p \leqslant 1\}$.

(b) If $\mathbf{B} = l_2$ and the distribution is Gaussian with independent components, the jth being $N(0, \sigma_j^2)$, then

$$K = \{(x_1, x_2, \ldots) \in l_2 \,|\, \Sigma(x_j/\sigma_j)^2 \leqslant 1\}.$$

(c) If $\mathbf{B} = C[0, 1]$ and we consider the distribution of the Brownian bridge, then

$$K = \left\{ x \in C[0, 1] \,\Big|\, x(t) = \int_0^t y(s)\, ds, \int_0^1 y^2(s)\, ds \leq 1, \int_0^1 y(s)\, ds = 0 \right\}.$$

(2) The careful reader will have noted that Theorem 2 does not quite reduce to Theorem 1 in one dimension. Indeed, it requires the information $\{E \min\{X_1, \ldots, X_n\}\}_{n=1}^\infty$ as well as $\{E \max\{X_1, \ldots, X_n\}\}_{n=1}^\infty$. In higher dimensions, this corresponds to considering both x^* and $-x^*$. It seems natural and inoffensive to retain this redundancy.

(3) One may ask whether Theorem 2 can be extended to random sets. That is, if $\mathbf{X}_1, \mathbf{X}_2, \ldots$ are iid random sets with $E\|\mathbf{X}_1\| < \infty$, is their common distribution determined by $\{E \operatorname{conv}\{\mathbf{X}_1 \cup \mathbf{X}_2 \cup \ldots \cup \mathbf{X}_n\}\}_{n=1}^\infty$? The answer is no as can be seen on the line: let $\mathbf{X}_i$ by a random interval $[\alpha_i, \beta_i]$. Then $E \operatorname{conv}\{\mathbf{X}_1 \cup \ldots \cup \mathbf{X}_n\} = [E \min\{\alpha_1, \ldots, \alpha_n\}, E \max\{\beta_1, \ldots, \beta_n\}]$. No information on the joint distribution of the end-points of the random intervals is carried by these sets.

(4) In finite dimensions, when the probability measure has, for instance, a density supported by a compact, convex set K_0, then the sequence of moment bodies lies in K_0 and expands out to the boundary of K_0. This suggests a novel method for approximating convex surfaces which will be treated elsewhere (Vitale, 1986).

(5) Variants of Theorem 2 can be derived in ways similar to the univariate case. For instance, not all moment bodies are needed for characterization. It is enough to have those for sample sizes $\{n_j\}$ satisfying Muntz's condition,

$$\sum_{j=1}^\infty \frac{1}{n_j} = \infty$$

(Huang, 1975).

References

Artstein, Z. (1974) 'On the Calculus of Closed Set-Valued Functions', *Indiana Univ. Math. J.* **24**, 433–441.

Artstein, Z. and Hansen, J. (1983) 'Convexification in Limit Laws of Random Sets in Banach Spaces', *Ann. Probab.* **13**, 307–309.

Artstein, Z. and Burns, J. (1974) 'Integration of Compact Set-Valued Functions', *Pacific J. Math.* **58**, 297–307.

Artstein, Z. and Vitale, R. A. (1975) 'A Strong Law of Large Numbers for Random Compact Sets', *Ann. Probab.* **3**, 879–882.

Aumann, R. J. (1965) 'Integrals of Set-Valued Functions', *J. Math. Anal. Appl.* **12**, 1–12.

Cameron, R. H. and Martin, W. T. (1944) 'Transformation of Wiener Integrals under Translations', *Ann. Math.* **45**, 386–396.

Debreu, G. (1967) 'Integration of Correspondences', *Proc. Fifth Berkeley Symposium on Math. Statist. and Probab.* Vol. 2, Univ. of California Press, pp. 351–372.

Eddy, W. F. (1980) 'The Distribution of the Convex Hull of a Gaussian Sample', *J. Appl. Probab.* **17**, 686–695.

Eddy, W. F. and Gale, J. D. (1981) 'The Convex Hull of a Spherically Symmetric Sample', *Adv. Appl. Probab.* **13**, 751–763.

Fisher, L. (1971) 'Limiting Convex Hulls of Samples: Theory and Function Space Examples', *Z. Wahrsch. Verw. Gebiete* **18**, 281–297.

Galambos, J. and Kotz, S. (1978) *Characterization of Probability Distributions*, Lecture Notes in Mathematics No. 675. Springer-Verlag, New York.

Giné, E. and Hahn, M. G. (1985) 'Characterization and Domains of Attraction of p-Stable Random Compact Sets', *Ann. Probab.* **13**, 447–468.

Gross, L. (1965) 'Abstract Wiener Spaces', *Proc. Fifth Berkeley Symp. Math. Statist. and Probab.*, Univ. of Calif. Press, pp. 31–42.

Hoeffding, W. (1951) '"Optimum" Non-Parametric Tests', *Proc. Second Berkeley Symp. Math. Statist. Probab.*, Univ. of Calif. Press, pp. 83–92.

Hoeffding, W. (1953) 'On the Distribution of the Expected Values of the Order Statistics', *Ann. Math. Statist.* **24**, 93–100.

Huang, J. S. (1975) 'Characterization of Distributions by the Expected Values of the Order Statistics', *Ann. Inst. Statist. Math.* **27**, 87–93.

Segal, I. E. (1956) 'Tensor Algebras over Hilbert Spaces', *Trans. Amer. Math. Soc.* **81**, 106–134.

Segal, I. E. (1958) 'Distributions in Hilbert Spaces and Canonical Systems of Operators', *Trans. Amer. Math. Soc.* **88**, 12–41.

Strassen, V. (1964) 'An Invariance Principle for the Law of the Iterated Logarithm', *Z. Wahrsch. Verw. Gebiete* **3**, 211–226.

Terry, M. E. (1952) 'Some Rank Order Tests which are Most Powerful Against Specific Parametric Alternatives', *Ann. Math. Statist.* **23**, 346–366.

Vitale, R. A. (1985) 'The Steiner Point in Infinite Dimensions', *Israel J. Math.* **52**, 245–250.

Vitale, R. A. (1986) 'Convex Bodies; Approximation from Within by Expected Convex Hulls' (in preparation).

Weil, W. (1982) 'An Application of the Central Limit Theorem for Banach-Space-Valued Random Variables to the Theory of Random Sets', *Z. Wahrsch. Verw. Gebiete* **60**, 203–208.

Acta Applicandae Mathematicae 9 (1987) 103–136
© 1987 by D. Reidel Publishing Company

Point Processes of Cylinders, Particles and Flats

WOLFGANG WEIL
Mathematisches Institut II, Universität Karlsruhe, Englerstrasse 2, 7500 Karlsruhe 1, F.R.G.

(Received: 9 April 1986; revised: 30 September 1986)

Abstract. Point processes X of cylinders, compact sets (particles), or flats in $\mathbf{R}^d$ are mathematical models for fields of sets as they occur, e.g., in practical problems of image analysis and stereology. For the estimation of geometric quantities of such fields, mean value formulas for X are important. By a systematic approach, integral geometric formulas for curvature measures are transformed into density formulas for geometric point processes. In particular, a number of results which are known for stationary and isotropic Poisson processes of convex sets are generalized to nonisotropic processes, to non-Poissonian processes, and to processes of nonconvex sets. The integral geometric background (including recent results from translative integral geometry), the fundamentals of geometric point processes, and the resulting density formulas are presented in detail. Generalizations of the theory and applications in image analysis and stereology are mentioned shortly.

AMS Subject classification (1980). Primary 60D05; secondary 60G55, 60–02, 52A22, 53C65, 62P99.

Key words. Geometric point process, image analysis, stereology, quermass density, integral geometry, curvature measure.

1. Introduction

In many practical situations problems of the following type occur. There is a collection of sets in $\mathbf{R}^2$ or $\mathbf{R}^3$, from which a transformed image is observed, e.g., a section, a projection, or the part within a 'sampling window'. Typically, the sets are compact particles or flats (lines, planes) or cylinders. In practice they may be pores in a porous medium, tubules in a tissue, fibres in a fabric, etc. The problem is to estimate geometric quantities of the entire collection based on information from the observed image. This is a classical stereological situation to which the statistical theory, developed on the basis of integral geometry by Davy and Miles (see Davy (1978), Miles (1978), or Weil (1983a), for surveys), may be applied. However, if the field $\mathscr{F}$ of sets is significantly extended (e.g., compared to the bounded sampling window) and if the interest is mainly in mean values (mean surface area of particles per unit volume, mean thickness of cylinders per unit volume, mean number of lines per unit area, etc.), $\mathscr{F}$ may be also considered as the outcome $\mathscr{F} = X(\omega)$ of a random process X, a geometric point process. In particular, if X can be assumed to have certain invariance properties (stationarity, isotropy) of its distribution, this point of view has a number of advantages. First of all, sections and sampling windows need not be chosen randomly in this case. Moreover, special point process models allow a statistical analysis which is quite complicated in the deterministic case. Finally, various geometric point processes can be generated by computer programs and, hence, many numerical results can be obtained by simulation.

The basic point process model used in the literature is that of a stationary and isotropic Poisson process of convex sets (particles, cylinders, flats). Formulas for such point processes are collected in Matheron (1975) and Davy (1978). More recently it has been shown that many results can be obtained under weaker assumptions, for example, for stationary and nonisotropic processes, for non-Poissonian processes, and for processes of nonconvex sets. Such results are found here and there in the literature and are obtained by quite different methods under varying assumptions. Matheron, in his book (Matheron (1975)), already emphasized the role of the curvature measures of a set as a basic notion in stochastic geometry. This opinion was supported by results in Weil (1983b), (1984), Zähle (1986). It turns out that integral geometric formulas for curvature measures can be transformed easily into density formulas for geometric point processes from which many (but, of course, not all) stereological results for mean values come out as special cases.

In the following this method is exploited systematically in d-dimensional space $\mathbf{R}^d$ and the resulting formulas are collected. Most of them are known but some are new and the others are shown to be true under quite general assumptions. In order to allow a unified approach to all the results the underlying class of sets is of importance. We base our considerations upon the 'convex ring' $\mathscr{R}_d$, a choice which is essential in view of the integral formulas for curvature measures which are used. Since compact sets can be approximated by sets from the convex ring, formulas given in the literature for more general sets (e.g., fibres, surfaces, etc.) may be obtained immediately from corresponding results on $\mathscr{R}_d$. This is sufficient for applications but it should be mentioned that the approximation of one set class by another such that geometric functionals converge is, in general, a difficult mathematical problem.

2. Integral Formulas for Curvature Measures

We need the following classes of sets in $\mathbf{R}^d$: $\mathscr{K}_d$ the class of convex bodies (compact convex subsets of $\mathbf{R}^d$), $\mathscr{R}_d$ the convex ring (finite unions of convex bodies), $\mathscr{L}_q^d$ the set of q-dimensional linear subspaces of $\mathbf{R}^d$, and $\mathscr{E}_q^d$ the set of q-flats (q-dimensional affine subspaces) in $\mathbf{R}^d$. For a subspace $L \in \mathscr{L}_q^d$ and a set K in the orthogonal space $L^\perp$, the vector sum $K + L$ is called a *cylinder* with basis K. Let $\mathscr{Z}_q(K)$ be the set of all cylinders (properly) congruent to $K + L$. The Lebesgue measure in $\mathbf{R}^d$ is denoted by λ_d. Correspondingly, λ_E is the Lebesgue measure in $E \in \mathscr{E}_q^d$ and, for a face F of a polytope or a cylinder, λ_F is the appropriate Lebesgue measure restricted to F. For $L, L' \in \mathscr{L}_q^d$, $[L, L']$ is the volume (of appropriate dimension) of the parallelepiped spanned by the following vectors. First choose an orthonormal basis in $L \cap L'$, then extend it to an orthonormal basis in L and an orthonormal basis in L'.

For faces F, F' of convex polytopes K, K' (or cylinders with polytopal basis) we define the angle $\gamma(F, F', K, K')$ as the external angle of $K \cap (K' + x)$ at the face $F \cap (F' + x)$, where $x \in \mathbf{R}^d$ is such that the relative interior of F meets the relative interior of $F' + x$. This definition does not depend on the choice of x as long as $F \cap (F' + x)$ has dimension at most $\max\{0, \dim F + \dim F' - d\}$ for all x. If this is the case for all faces, we say that K and K' are in *general relative position*.

For $K \in \mathcal{K}_d$ the *curvature measures* $\Psi_0(K, \cdot), \ldots, \Psi_d(K, \cdot)$ can be introduced by the local Steiner formula

$$\lambda_d(A_\epsilon(K, \beta)) = \sum_{j=0}^{d} \epsilon^{d-j} \kappa_{d-j} \Psi_j(K, \beta). \tag{2.1}$$

Here, $\beta \subset \mathbf{R}^d$ is a Borel set, κ_k is the volume of the k-dimensional unit ball, $\epsilon > 0$, and $A_\epsilon(K, \beta)$ is a local parallel set (the set of all $x \in \mathbf{R}^d$ such that the metric projection $\mathrm{proj}_K(x)$ of x onto K obeys $\mathrm{proj}_K(x) \in \beta$ and $\|x - \mathrm{proj}_K(x)\| \leq \epsilon$). An important property of the curvature measures is that they depend additively on K. An additive extension to sets $K \in \mathcal{R}_d$ was given by Schneider (1980a). For $K = \cup_{i=1}^{n} K_i$, $K_i \in \mathcal{K}_d$, this extension $\Psi_j(K, \cdot)$ fulfills

$$\Psi_j(K, \cdot) = \sum_{i=1}^{n} \Psi_j(K_i, \cdot) - \sum_{1 \leq i_1 < i_2 \leq n} \Psi_j(K_{i_1} \cap K_{i_2}, \cdot) + $$
$$- \cdots + (-1)^{n+1} \Psi_j(K_1 \cap \cdots \cap K_n, \cdot). \tag{2.2}$$

Equation (2.2) cannot be used as a definition of $\Psi_j(K, \cdot)$ since the right side may depend on the choice of the sets $K_1, \ldots, K_n$. An alternative approach was given by Schneider (1980a) who showed that there is a generalization of (2.1) to sets K in $\mathcal{R}_d$. For properties of the curvature measures, we refer to the surveys Schneider (1979) and Weil (1983a). We mention only that for j-sets $K \in \mathcal{R}_d$, i.e., sets which are the union of convex bodies $K_1, \ldots, K_n$ of dimension at most j, the curvature measure $\Psi_j(K, \cdot)$ equals the j-dimensional Hausdorff measure on K. Since the curvature measures are locally defined, they extend immediately to unbounded sets, which are locally finite countable unions of convex bodies. In particular, this holds for cylinders Z. More precisely, for a cylinder Z, a convex body K', and a Borel set β in the interior $\mathrm{int}\, K'$ of K', we have

$$\Psi_j(Z, \beta) = \Psi_j(Z \cap K', \beta).$$

For stereological applications, the total measures $V_j(K) = \Psi_j(K, \mathbf{R}^d)$ are important. $V_j(K)$ is proportional to the $(d-j)$th quermass integral of K, in particular $V_d(K)$ is the volume, $V_{d-1}(K)$ is half the surface area, $V_{d-2}(K)$ is proportional to the integral mean curvature, $\ldots$, $V_1(K)$ is proportional to the (additively extended) mean width, and $V_0(K)$ is the Euler–Poincaré characteristic of $K \in \mathcal{R}_d$. It is convenient to define $V_j(Z)$ for a cylinder $Z = K + L$ not as the total curvature measure but as $V_{j-q}(K)$, $j = q, \ldots, d$.

The basic integral geometric result for curvature measures is the following translation formula for cylinders (Schneider and Weil (1986)). Let $L \in \mathcal{L}_q^d$, $K' \in \mathcal{R}_d$ with $K' \subset L^\perp$, $K \in \mathcal{R}_d$, and Borel sets $\beta \subset \mathbf{R}^d$, $\beta' \subset L^\perp$ be given. Then, for $j \in \{0, \ldots, d\}$,

$$
\int_{L^\perp} \Psi_j(K \cap (K' + L + x), \beta \cap (\beta' + L + x)) \, d\lambda_{L^\perp}(x)
$$

$$
= \Psi_j(K, \beta) \Psi_{d-q}(K', \beta') + \sum_{k=j+1}^{d-1} \varphi_k^{(j)}(K, K' + L, \beta \times \beta') + \tag{2.3}
$$

$$
+ \Psi_d(K, \beta) \Psi_{j-q}(K', \beta')
$$

(curvature measures with index $m < 0$ are defined to be zero). $\varphi_k^{(j)}(K, K' + L, \cdot)$ is a (signed) Borel measure on $\mathbf{R}^d \times \mathbf{R}^d$ which depends in an additive and measurable way on K, K' (and L). For $k > d + j - q$, $\varphi_k^{(j)}(K, K' + L, \cdot)$ vanishes, hence the summation in (2.3) really extends only to $d + j - q$, if $j < q$. The measures $\varphi_k^{(j)}(K, K' + L, \cdot)$ with $k \leq d + j - q$, are concentrated on bd $K \times$ bd K' (here bd denotes the boundary). For $k \leq d + j - q$, $\varphi_k^{(j)}(K, K' + L, \cdot)$ is explicitly known if K, K' are convex polytopes and K, $K' + L$ are in general relative position. Then we have

$$
\varphi_k^{(j)}(K, K' + L, \cdot)
$$

$$
= \sum_{F \in \mathcal{F}_k(K)} \sum_{F' \in \mathcal{F}_{d+j-q-k}(K')} \gamma(F, F' + L, K, K' + L) \times \tag{2.4}
$$

$$
\times [L(F), L(F') + L] \lambda_F \otimes \lambda_{F'}.
$$

Here, $\mathcal{F}_m(P)$ is the set of m-dimensional faces of the polytope P and $L(F) \in \mathcal{L}_m^d$ is the subspace parallel to $F \in \mathcal{F}_m(P)$. Two special cases of (2.3) are of interest. First, if $L = \{0\}$, we get

$$
\int_{\mathbf{R}^d} \Psi_j(K \cap (K' + x), \beta \cap (\beta' + x)) \, d\lambda_d(x)
$$

$$
= \Psi_j(K, \beta) \Psi_d(K', \beta') + \tag{2.5}
$$

$$
+ \sum_{k=j+1}^{d-1} \varphi_k^{(j)}(K, K', \beta \times \beta') + \Psi_d(K, \beta) \Psi_j(K', \beta').
$$

Next, if $K' = \{0\}$, the measure $\varphi_k^{(j)}(K, K' + L, \cdot)$ is concentrated on bd $K \times \{0\}$, hence it may be viewed as a measure on $\mathbf{R}^d$ (concentrated on bd K). Since $\varphi_k^{(j)}(K, K' + L, \cdot)$ depends homogeneously of degree $d + j - q - k$ on K' it must vanish as long as $d + j - q - k \neq 0$. Hence, from (2.3) it follows that

$$
\int_{L^\perp} \Psi_j(K \cap (L + x), \beta \cap (L + x)) \, d\lambda_{L^\perp}(x)
$$

$$
= \varphi_{d+j-q}^{(j)}(K, L, \beta). \tag{2.6}
$$

An important property of the measure $\varphi_k^{(j)}(K, K' + L, \cdot)$ concerns its rotation integral. If SO_d denotes the rotation group with invariant measure ν, $\nu(SO_d) = 1$, then

$$\int_{SO_d} \varphi_k^{(j)}(K, \theta(K' + L), \beta \times \theta\beta') \, d\nu(\theta)$$
$$= \alpha_{djk} \Psi_k(K, \beta) \Psi_{d+j-q-k}(K', \beta'). \tag{2.7}$$

Combining (2.7) with (2.3), we get the kinematic formula for cylinders (Schneider 1980b)

$$\int_{SO_d} \int_{L^\perp} \Psi_j(K \cap \theta(K' + L + x), \beta \cap \theta(\beta' + L + x)) \, d\lambda_{L^\perp}(x) \, d\nu(\theta)$$
$$= \sum_{k=j}^{d} \alpha_{djk} \Psi_k(K, \beta) \Psi_{d+j-q-k}(K', \beta'). \tag{2.8}$$

Again as special cases, (2.8) contains the *principal kinematic formula* and the *Crofton formula* for curvature measures. Instead of using double integrals we consider here the group G_d of rigid motions with invariant measure μ and the space $\mathscr{E}_q^d$ with invariant measure μ_q. Then

$$\int_{G_d} \Psi_j(K \cap gK', \beta \cap g\beta') \, d\mu(g)$$
$$= \sum_{k=j}^{d} \alpha_{djk} \Psi_k(K, \beta) \Psi_{d+j-k}(K', \beta') \tag{2.9}$$

and

$$\int_{\mathscr{E}_q^d} \Psi_j(K \cap E, \beta \cap E) \, d\mu_q(E) = \alpha_{djq} \Psi_{d+j-q}(K, \beta). \tag{2.10}$$

The coefficients α_{djk} occurring in (2.7) to (2.10) are given by

$$\alpha_{djk} = \frac{\binom{k}{j} \kappa_k \kappa_{d+j-k}}{\binom{d}{k-j} \kappa_j \kappa_d}.$$

A variant of (2.3) and (2.8) concerns the case of projected thick sections. For a cylinder $K' + L$ denote by $A | L$ the projection of the set $A \subset \mathbf{R}^d$ onto the direction space L of $K' + L$. Then, we have for $K, K' \in \mathscr{K}_d$

$$\int_{L^\perp} \Psi_j([K \cap (K' + L + x)] | L, [\beta \cap (\beta' + L + x)] | L) \, d\lambda_{L^\perp}(x)$$
$$= \varphi_{d+j-q}^{(j)}(K + (-K'), L, \beta + (-\beta')), \tag{2.11}$$

where

$$-K' = \{-x \mid x \in K'\}.$$

The analog of (2.8) would be

$$\int_{SO_d} \int_{L^\perp} \Psi_j([K \cap \theta(K'+L+x)] | \theta L, [\beta \cap \theta(\beta'+L+x)] | \theta L) \, d\lambda_{L^\perp}(x) \, d\nu(\theta)$$

$$= \sum_{k=j}^{d+j-q} \gamma_{djkq} \Psi_k(K, \beta) \Psi_{d+j-q-k}(K', \beta') \tag{2.12}$$

with

$$\gamma_{djkq} = \frac{\binom{q}{j} \kappa_q \kappa_k \kappa_{d-k}}{\binom{d}{d-k} \kappa_{q-j} \kappa_d \kappa_j}.$$

So far, (2.12) is known only for the quermassintegrals, i.e., for $\beta = \mathbf{R}^d$, $\beta' = L^\perp$ (see Schneider (1981)). The two formulas can be extended to sets $K \in \mathcal{R}_d$ which are unions of convex bodies $K_1, \ldots, K_n$, provided the different parts $K_1, \ldots, K_n$ of K can be distinguished in the projection and the integrands are modified appropriately. For practical applications, the following case is important: K is a 1-set, i.e., $K_1, \ldots, K_n$ are line segments, no two of which have more than one point in common. Then, (2.12) holds for the quermassintegral V_1 without further modification, if $d \geq 3$.

Since

$$\Psi_j(K, \mathbf{R}^d) = V_j(K), \quad j = 0, \ldots, d, K \in \mathcal{R}_d,$$

the integral formulas (2.3) to (2.12) contain, as special cases, integral formulas for quermassintegrals. It is therefore of interest to obtain more explicit expressions for the total measures $\Phi_k^{(j)}(K, Z) = \varphi_k^{(j)}(K, Z, \mathbf{R}^d \times \mathbf{R}^d)$ (resp. $\Phi_k^{(j)}(K, L) = \varphi_k^{(j)}(K, L, \mathbf{R}^d)$). We have (Goodey and Weil (1986))

$$\Phi_{d+j-q}^{(j)}(K, K'+L) = \frac{\binom{d}{q-j}}{\kappa_{q-j}} V(\underbrace{K, \ldots, K}_{d+j-q}, \underbrace{B_L, \ldots, B_L}_{q-j}), \tag{2.13}$$

where $V(K_1, \ldots, K_d)$ denotes the mixed volume of $K_1, \ldots, K_d$ and B_L is the q-dimensional unit ball in L. Also,

$$\Phi_k^{(0)}(K, K'+L)$$

$$= \frac{\binom{d}{q}\binom{d-q}{k}}{\kappa_q} V(\underbrace{K, \ldots, K}_{k}, \underbrace{-K', \ldots, -K'}_{d-q-k}, \underbrace{B_L, \ldots, B_L}_{q}), \tag{2.14}$$

$k = 0, \ldots, d-q$, and

$$\Phi_k^{(j)}(K, K' + L)$$

$$= \frac{\binom{d}{q}\binom{d-q}{k-j}\binom{d}{j}\kappa_d}{\kappa_q \kappa_{d-j} \kappa_j} \times \qquad (2.15)$$

$$\times \int_{\mathscr{E}_{d-j}^d} V(\underbrace{K \cap E, \ldots, K \cap E}_{k-j}, \underbrace{-K', \ldots, -K'}_{d+j-q-k}, \underbrace{B_L, \ldots, B_L}_{q}) \, d\mu_{d-j}(E),$$

$k = j+1, \ldots, \min(d-1, d+j-q)$. Because of the use of mixed volumes, these results first hold for convex bodies K, K', but they can be extended to sets K, $K' \in \mathscr{R}^d$ by additivity. The extension is possible since the mixed volumes in (2.13), (2.14), and (2.15) depend additively and continuously on the sets K, $K' \in \mathscr{K}_d$ (see the article of McMullen and Schneider (1983), for more details).

Of special interest is the case where K is an r-dimensional convex body, $0 \leq r \leq d-1$. Then using (2.15) we see that $\Phi_k^{(j)}(K, K'+L) = 0$ if $k > r$. Thus, if we assume $0 < r - j \leq d - q$, the last summand in (2.3) (for $\beta \times \beta' = \mathbb{R}^d \times L^\perp$) is $\Phi_r^{(j)}(K, K'+L)$. Here, formula (2.15) can be simplified (Goodey and Weil (1986)):

$$\Phi_r^{(j)}(K, K' + L)$$

$$= \frac{\binom{d}{q}\binom{d-q}{r-j}}{\kappa_{r-j}\kappa_q} V_r(K) V(\underbrace{K', \ldots, K'}_{d+j-r-q}, \underbrace{B_L, \ldots, B_L}_{q}, \underbrace{B_M, \ldots, B_M}_{r-j}) \qquad (2.16)$$

($M \in \mathscr{L}_r^d$ is the subspace parallel to the affine hull of K).

3. Geometric Point Processes

Let $\mathscr{Z}_q$ be the set of all cylinders in $\mathbb{R}^d$ with q-dimensional direction space and basis in $\mathscr{R}_d$. For some fixed $L \in \mathscr{L}_q^d$, $\mathscr{Z}_q$ is the union of all sets $\mathscr{Z}_q(K)$, $K \subset L^\perp$, $K \in \mathscr{R}_d$. A *point process* X on $\mathscr{Z}_q$ is given by a probability measure P_X, the distribution of X, on the set $\mathscr{M}(\mathscr{Z}_q)$ of locally finite counting measures on $\mathscr{Z}_q$ supplied with its usual σ-algebra. Since $\mathscr{Z}_q$ is a measurable subset of the set $\mathscr{F}$ of all closed sets in $\mathbb{R}^d$, this definition of X fits into the general theory of point processes on $\mathscr{F}$ (see Matheron (1975) for details). Contrary to the definition of point processes in general spaces (see e.g., Neveu (1977)), we call a measure φ on $\mathscr{Z}_q$ *locally finite*, if

$$\varphi(\{Z \in \mathscr{Z}_q \mid \operatorname{conv} Z \cap K \neq \emptyset\}) < \infty$$

for all $K \in \mathscr{R}_d$. Alternatively, we consider a point process X as a measurable mapping from some abstract probability space $(\Omega, \mathscr{A}, P)$ into $\mathscr{M}(\mathscr{Z}_q)$. We therefore write $X(\omega)$ for a realization of X. Since we only work with simple point processes, we may interpret a counting measure η on $\mathscr{Z}_q$ also as a collection of cylinders. This allows us to write $Z \in X(\omega)$ or $Z \in X$.

The point process X on $\mathscr{Z}_q$ is *stationary* if P_X is translation invariant, it is *isotropic* if P_X is invariant w.r.t. rotations. The *intensity measure* Θ of X is defined by

$$\Theta(\mathscr{B}) = \mathbb{E}\,X(\mathscr{B})$$

for any Borel set $\mathscr{B} \subset \mathscr{Z}_q$. We assume throughout that Θ is locally finite. If X is stationary (isotropic), Θ has the same invariance properties but the converse is obviously false. X is called *weakly stationary* (*weakly isotropic*) if Θ is stationary (isotropic).

The following decomposition is important. Let $s: \mathscr{R}_d \to \mathbf{R}^d$ be a measurable mapping which associates to each set $K \in \mathscr{R}_d$ a center $s(K)$ in a motion covariant way. Familiar choices of such centers are the midpoint of the circumsphere or the Steiner point (the usual centroid does not exist for all sets in $\mathscr{R}_d$).

Let $\mathscr{Z}_q^0$ be the set of all cylinders $Z = K + L$ such that $L \in \mathscr{L}_q^d$, $K \in \mathscr{R}_d$, $K \subset L^\perp$, and $s(K) = 0$, and let

$$\tilde{\mathscr{Z}}_q = \{(x, Z) \mid Z \in \mathscr{Z}_q^0, Z = K + L, x \in L^\perp\}.$$

Then $i: (x, Z) \mapsto Z + x$ is an isomorphism of $\tilde{\mathscr{Z}}_q$ onto $\mathscr{Z}_q$. Assume now that X is weakly stationary. Then there exists a number $\gamma \geq 0$ and a probability measure P_0 on $\mathscr{Z}_q^0$ such that

$$i \circ \Theta(A \times C) = \gamma \int_C \lambda_{L(Z)^\perp}(A)\,dP_0(Z), \tag{3.1}$$

for Borel sets $A \subset \mathbf{R}^d$, $C \subset \mathscr{Z}_q^0$. Here, $L(Z)$ is the direction space of the cylinder Z. We call γ the *intensity* and P_0 the *shape distribution of* the point process X. Moreover, X is weakly isotropic, if and only if P_0 is rotation invariant. Frequently, we use the abbreviation $P_0(f)$ for the integral $\int_{\mathscr{Z}_q^0} f(Z)\,dP_0(Z)$. If X is a process of (convex) particles, P_0 is concentrated on $\mathscr{R}_d^0$ or $\mathscr{K}_d^0$, the set of particles K in $\mathscr{R}_d$ resp. $\mathscr{K}_d$ with $s(K) = 0$. If X is a process of q-flats, P_0 is concentrated on $\mathscr{L}_q^d$.

An important tool for the following is the Campbell theorem. In view of (3.1), it states that for every measurable and Θ-integrable function f on $\mathscr{Z}_q$

$$\mathbb{E} \sum_{Z \in X} f(Z) = \gamma \int_{\mathscr{Z}_q^0} \int_{L^\perp} f(K + L + x)\,d\lambda_{L^\perp}(x)\,dP_0(K + L). \tag{3.2}$$

We first use this result to give an interpretation of γ. For $K \in \mathscr{K}_d$ let

$$\mathscr{Z}_q(K) = \{Z \in \mathscr{Z}_q \mid Z \cap K \neq \emptyset\}$$

and let $1_{\mathscr{Z}_q(K)}(\cdot)$ be the indicator function of $\mathscr{Z}_q(K)$. Let B denote the unit ball in $\mathbf{R}^d$.

THEOREM 3.1. *We have*

$$\gamma = \lim_{r \to \infty} \frac{1}{r^{d-q} \kappa_{d-q}} \mathsf{E} \sum_{Z \in X} 1_{\mathscr{X}_q(rB)}(Z).$$

Proof. From (3.2) we get

$$\mathsf{E} \sum_{Z \in X} 1_{\mathscr{X}_q(rB)}(Z)$$

$$= \gamma \int_{\mathscr{X}_q^0} \int_{L^{\perp}} 1_{\mathscr{X}_q(rB)}(K + L + x) \, d\lambda_{L^{\perp}}(x) \, dP_0(K + L)$$

$$= \gamma \int_{\mathscr{X}_q^0} \lambda_{L^{\perp}}[(rB \cap L^{\perp}) + K] \, dP_0(K + L).$$

In the same way, we get from the Steiner formula (2.1)

$$\Theta(\{Z \in \mathscr{X}_q \mid \operatorname{conv} Z \cap rB \neq \emptyset\})$$

$$= \gamma \int_{\mathscr{X}_q^0} \lambda_{L^{\perp}}[(rB \cap L^{\perp}) + \operatorname{conv} K] \, dP_0(K + L)$$

$$= \sum_{j=0}^{d-q} r^{d-q-j} \kappa_{d-q-j} \int_{\mathscr{X}_q^0} V_j(\operatorname{conv} K) \, dP_0(K + L).$$

Since

$$\left| \frac{1}{r^{d-q} \kappa_{d-q}} \lambda_{L^{\perp}}[(rB \cap L^{\perp}) + K] \right|$$

$$\leq \sum_{j=0}^{d-q} r^{-j} \frac{\kappa_{d-q-j}}{\kappa_{d-q}} V_j(\operatorname{conv} K)$$

and since Θ is locally finite, the integrand in the following integral is uniformly integrable and we get

$$\lim_{r \to \infty} \frac{1}{r^{d-q} \kappa_{d-q}} \mathsf{E} \sum_{Z \in X} 1_{\mathscr{X}_q(rB)}(Z)$$

$$= \gamma \int_{\mathscr{X}_q^0} \lim_{r \to \infty} \frac{1}{r^{d-q} \kappa_{d-q}} \lambda_{L^{\perp}}[(rB \cap L^{\perp}) + K] \, dP_0(K + L)$$

$$= \gamma.$$

If in Theorem 3.1 the indicator function $1_{\mathscr{X}_q(rB)}$ is replaced by $1_{\mathscr{X}_q(rB)} \cdot f$, where f is a translation invariant function on $\mathscr{X}_q$ such that

$$\int_{\mathscr{X}_q^0} V_i(\operatorname{conv} Z)|f(Z)| \, dP_0(Z) < \infty$$

for $i = q, \ldots, d$, then the same arguments hold true and we get

$$\lim_{r \to \infty} \frac{1}{r^{d-q} \kappa_{d-q}} \mathbb{E} \sum_{Z \in X} 1_{\mathcal{X}_q(rB)}(Z) \cdot f(Z)$$
$$= \gamma \cdot P_0(f).$$

We are especially interested in the case $f = V_j$.

THEOREM 3.2. *Suppose* $\int_{\mathcal{X}_q^0} V_i(\text{conv } Z) |V_j(Z)| \, dP_0(Z) < \infty$ *for* $i, j = q, \ldots, d$. *Then*

$$D_j(X) := \lim_{r \to \infty} \frac{1}{r^{d-q} \kappa_{d-q}} \mathbb{E} \sum_{Z \in X} 1_{\mathcal{X}_q(rB)}(Z) V_j(Z)$$

exists and fulfills

$$D_j(X) = \gamma P_0(V_j), \quad j = q, \ldots, d.$$

We call $D_j(X)$ the *jth quermass density* of the point process X. If the cylinders Z of the process are all simply connected, then $V_q(Z) = 1$ and, hence, $D_q(X) = \gamma$.

Results similar to Theorems 3.1 and 3.2 hold if the unit ball B is replaced by an arbitrary convex body K with dimension at least $d - q$, but the normalizing constant κ_{d-q} then looks more complicated. It has a simple form, namely $V_{d-q}(K)$, if X is weakly isotropic. Two special cases are of interest. If the cylinders are a.s. q-flats, i.e., if P_0 is concentrated on those $Z = K + L \in \mathcal{X}_q^0$ with $K = \{0\}$, and if X is weakly isotropic, then γ gives the mean number of flats of the process X per unit $(d - q)$-dimensional volume. A better mathematical basis of this interpretation of γ will be given later. Second, if $q = 0$, then X is a process of particles. In this case we have

$$\gamma = \lim_{r \to \infty} \frac{1}{V_d(rK_0)} \mathbb{E} \sum_{K \in X} 1_{\mathcal{R}_d(rK_0)}(K)$$

and

$$D_j(X) = \gamma P_0(V_j) = \lim_{r \to \infty} \frac{1}{V_d(rK_0)} \mathbb{E} \sum_{K \in X} 1_{\mathcal{R}_d(rK_0)}(K) V_j(K)$$

without requiring that X is isotropic. Here $K_0 \in \mathcal{K}_d$ is arbitrary (but with inner points) and $\mathcal{R}_d(rK_0) = \{K \in \mathcal{R}_d \mid K \cap rK_0 \neq \emptyset\}$. The proof is similar to that of Theorem 3.1 and Theorem 3.2 but, instead of the Steiner formula, mixed volumes have to be used.

4. Formulas for Quermass Densities

With the help of (3.2) it is now easy to apply integral formulas for curvature measures to point processes. Let us start with (2.3). Our assumptions are the following. We consider a point process X of cylinders (i.e., a point process on $\mathcal{X}_q$ with $q \in \{0, \ldots, d-1\}$) which is weakly stationary (and fulfills the appropriate integrability conditions of Section 3) and a set $K_0 \in \mathcal{R}_d$ which serves as a sampling window. Moreover, we fix a Borel set β_0 (associated with K_0) and we assume that to each cylinder $Z = K + L$ a Borel set $\beta(Z) = \beta(K) + L$ is associated in a measurable and translation covariant way, i.e., we have

$$\beta(Z + x) = \beta(Z) + x$$

for all $x \in \mathbf{R}^d$. Then from (3.2) and (2.3) we get

$$\mathbb{E} \sum_{Z \in X} \Psi_j(K_0 \cap Z, \beta_0 \cap \beta(Z))$$

$$= \gamma \int_{\mathcal{X}_q^0} \int_{L^\perp} \Psi_j(K_0 \cap (K + L + x), \beta_0 \cap (\beta(K) + L + x)) \, d\lambda_{L^\perp}(x) \, dP_0(K + L)$$

$$= \gamma \Bigg[\Psi_j(K_0, \beta_0) \int_{\mathcal{X}_q^0} \Psi_{d-q}(K, \beta(K)) \, dP_0(K + L) + \qquad (4.1)$$

$$+ \sum_{k=j+1}^{d-1} \int_{\mathcal{X}_q^0} \varphi_k^{(j)}(K_0, Z, \beta_0 \times \beta(Z)) \, dP_0(Z) +$$

$$+ \Psi_d(K_0, \beta_0) \int_{\mathcal{X}_q^0} \Psi_{j-q}(K, \beta(K)) \, dP_0(K + L) \Bigg],$$

for $j = 0, \ldots, d$.

Because of our notation, the last summand in (4.1) vanishes if $j < q$ and, in this case, the summation is from $j+1$ to $d+j-q$.

Let us examine some special cases of formula (4.1). First, if $\beta_0 = \beta(Z) = \mathbf{R}^d$, then we get

$$\mathbb{E} \sum_{Z \in X} V_j(K_0 \cap Z)$$

$$= \gamma V_j(K_0) P_0(V_d) + \qquad (4.2)$$

$$+ \gamma \sum_{k=j+1}^{d-1} P_0(\Phi_k^{(j)}(K_0, \cdot)) + \gamma V_d(K_0) P_0(V_j).$$

The right-hand side can be expressed by means of mixed volumes using (2.14) and (2.15). In view of the homogeneity properties, (4.2) implies the following interpretation of $D_j(X)$.

THEOREM 4.1. *We have*

$$D_j(X) = \lim_{r \to \infty} \frac{1}{V_d(rK_0)} \mathbb{E} \sum_{Z \in X} V_j(rK_0 \cap Z)$$

for all $K_0 \in \mathcal{R}_d$ with inner points, $j = q, \ldots, d$.

To give another interpretation of $D_j(X)$ let $\beta_0 \subset \mathrm{int} \, K_0$ and $\beta(Z) = \mathbf{R}^d$. Then

$\Psi_j(K_0, \beta_0) = 0$, for $j < d$, and $\varphi_k^{(j)}(K_0, Z, \beta_0 \times \mathbf{R}^d) = 0$, for $k < d$. Hence, (4.1) implies

$$\mathsf{E} \sum_{Z \in X} \Psi_j(K_0 \cap Z, \beta_0) = \gamma \Psi_d(K_0, \beta_0) P_0(V_j). \tag{4.3}$$

Since the curvature measures are locally defined and since we may choose to each bounded Borel set β_0 a set $K_0 \in \mathcal{R}_d$ with $\beta_0 \subset \operatorname{int} K_0$, this implies the following result.

THEOREM 4.2. *We have*

$$\mathsf{E} \sum_{Z \in X} \Psi_j(Z, \cdot) = D_j(X) \cdot \lambda_d,$$

for $j = q, \ldots, d$.

Whilst the equation

$$D_j(X) = \gamma P_0(V_j), \quad j = q, \ldots, d,$$

presents the most direct way to define the quermass density $D_j(X)$ for the point process X, the formulas in Theorems 3.2, 4.1, and 4.2 describe other possible approaches, namely approaches of 'ergodic' type and the random measure approach. As the results show, the different methods lead to the same quantity, namely $D_j(X)$. Similar considerations have been made for random sets and point processes of sets from other set classes (Weil, 1984; Weil and Wieacker, 1984; Zähle, 1986). Using Lemma 6 in Weil and Wieacker (1984) one also gets

$$D_j(X) = \mathsf{E} \sum_{Z \in X} [V_j(C_0 \cap Z) - V_j(\delta^+ C_0 \cap Z)]$$

$j = q, \ldots, d$, where C_0 is a unit cube and $\delta^+ C_0$ its 'upper right' boundary (see Weil and Wieacker (1984) for details).

If the point process X is ergodic, then the formulas in Theorems 3.2 and 4.1 hold for almost all realizations $X(\omega)$ (or in L^1) without the expectation sign. In particular, any Poisson process X is ergodic (see Nguyen and Zessin (1979) and Wieacker (1982) for more details).

As a consequence of Theorem 4.2 (or (4.3)) we get the following interesting formula:

$$\mathsf{E} \sum_{Z \in X} \Psi_j(Z, \operatorname{int} K_0) = V_d(K_0) D_j(X), \quad j = q, \ldots, d. \tag{4.4}$$

As other special cases of (4.1) we put $\beta_0 = \mathbf{R}^d$ and $\beta(Z) = \operatorname{int} Z$. Then

$$\mathsf{E} \sum_{Z \in X} \Psi_j(K_0, \operatorname{int} Z) = V_j(K_0) D_d(X). \tag{4.5}$$

For $\beta_0 = \text{bd } K_0$ and $\beta(Z) = \text{bd } Z$, we get

$$E \sum_{Z \in X} \Psi_j(K_0 \cap Z, \text{bd } K_0 \cap \text{bd } Z)$$

$$= \gamma \sum_{k=j+1}^{d-1} P_0(\Phi_k^{(j)}(K_0, \cdot)) \tag{4.6}$$

which can be expressed by mixed volumes again.

For weakly isotropic X, these formulas can be simplified further in view of the integral formula (2.8). In this case, (4.1) becomes

$$E \sum_{Z \in X} \Psi_j(K_0 \cap Z, \beta_0 \cap \beta(Z))$$

$$= \gamma \sum_{k=j}^{d} \alpha_{djk} \Psi_k(K_0, \beta_0) \int_{\mathscr{X}_q^0} \Psi_{d+j-q-k}(K, \beta(K)) \, dP_0(K+L), \tag{4.7}$$

$j = 0, \ldots, d$.

The corresponding changes in (4.2) and (4.6) are obvious.

For a process X of particles (i.e., $q = 0$), (4.2) can be written as

$$E \sum_{K \in X} V_j(K_0 \cap K)$$

$$= \gamma \Bigg[V_j(K_0) P_0(V_d) + \sum_{k=j+1}^{d-1} \frac{\binom{d}{k-j} \binom{d}{j} \kappa_d}{\kappa_{d-j} \kappa_j} \times$$

$$\times \int_{\mathscr{E}_{d-j}^d} P_0(V(\underbrace{-K_0 \cap E, \ldots, -K_0 \cap E}_{k-j}, \underbrace{\cdot, \ldots, \cdot}_{d+j-k})) \, d\mu_{d-j}(E) + \tag{4.8}$$

$$+ V_d(K_0) P_0(V_j) \Bigg].$$

For weakly isotropic X, this formula becomes

$$E \sum_{K \in X} V_j(K_0 \cap K) = \gamma \sum_{k=j}^{d} \alpha_{djk} V_k(K_0) P_0(V_{d+j-k}),$$

a result which was first obtained by Fava and Santaló (1978), (1979).

For another consequence of (4.2) let X be a process of flats. If $j = q$, we get a sharper form of Theorem 4.1.

THEOREM 4.3. *For a process X of q-flats we have*

$$\gamma = \frac{1}{V_d(K_0)} E \sum_{L \in X} V_q(K_0 \cap L).$$

Thus, the intensity of X can be also interpreted as the mean q-dimensional volume of the q-flats in X per unit volume of $\mathbf{R}^d$. If $j \neq q$, there is only one term on the right side of (4.2) which does not vanish, namely the summand $P_0(\Phi_{d+j-q}^{(j)}(K_0, \cdot))$. It can be expressed more explicitly with the help of (2.13).

THEOREM 4.4. *Let* $K_0 \in \mathcal{R}_d$ *and* X *be a point process of* q-*flats. Then*

$$\mathbb{E} \sum_{L \in X} V_j(K_0 \cap L)$$

$$= \gamma \frac{\binom{d}{q-j}}{\kappa_{q-j}} P_0(V(\underbrace{K_0, \ldots, K_0}_{d+j-q}, \underbrace{B_\bullet, \ldots, B_\bullet}_{q-j})).$$

In particular, if the affine hull of K_0 is $(d+j-q)$-dimensional (and $L_0 \in \mathcal{L}_{d+j-q}^d$ parallel to this affine hull), then

$$V(\underbrace{K_0, \ldots, K_0}_{d+j-q}, \underbrace{B_L, \ldots, B_L}_{q-j}) = V_{d+j-q}(K_0) \frac{\kappa_{q-j}}{\binom{d}{q-j}}[L_0, L]$$

and hence

$$\mathbb{E} \sum_{L \in X} V_j(K_0 \cap L) = \gamma \, V_{d+j-q}(K_0) \int_{\mathcal{L}_q^d} [L_0, L] \, dP_0(L). \tag{4.9}$$

If X is weakly isotropic, we use (4.7) instead of (4.1) and get a generalization of Theorem 4.3.

THEOREM 4.5. *Let* X *be a weakly isotropic point process of* q-*flats, let* $K_0 \in \mathcal{R}_d$, *and* $q \in \{0, \ldots, d-1\}$. *Then*

$$\gamma = \frac{1}{\alpha_{djq} V_{d+j-q}(K_0)} \sum_{L \in X} V_j(K_0 \cap L)$$

for $j = 0, \ldots, q$.

In particular, this result includes q different interpretations of γ as mean value per unit content if we take a $(d+j-q)$-dimensional set K_0.

5. Point Processes Induced on Lower Dimensional Flats

Let X be a point process of cylinders with the properties mentioned at the beginning of the last section (weakly stationary, etc.) and let E_r be a fixed r-flat, $r \in \{1, \ldots, d-1\}$. We consider the intersection process $X \cap E_r$ which consists of the sets $Z \cap E_r$, $Z \in X$. If $Z = K + L$ and if L and E_r are in general relative position, then $Z \cap E_r$ is either empty or a cylinder in $\mathcal{Z}_p$ with $p = \max\{0, q+r-d\}$. Let $\mathcal{Z}_{q,E_r}^0$ be the set of all cylinders $Z = K + L \in \mathcal{Z}_q^0$, for which L and E_r are not in general relative position. Then

$$\mu_r(\{E_r \mid P_0(\mathcal{Z}_{q,E_r}^0) > 0\}) = 0,$$

i.e., for μ_r-almost all E_r, almost all realizations $X(\omega)$ of X have the property that L and E_r are in general relative position for all $K + L \in X(\omega)$. Hence, for almost all E_r the intersection process $X \cap E_r$ is a process of cylinders in $\mathcal{Z}_p$. It may, however, be possible, even in this case that $Z \cap E_r = \emptyset$ for all $Z \in X$. This event occurs either with probability zero or with probability one. While the first possibility is, of course, of no significance, the second occurs, if and only if X is a process of q-flats and $q + r < d$. We may then include this 'empty' process $X \cap E_r$ into our considerations as a point process of cylinders with intensity zero.

After these explanations we can assume that $X \cap E_r$ is a point process of cylinders in $\mathcal{Z}_p$ (lying in E_r). It is obvious that $X \cap E_r$ has the same invariance properties as X. In particular, for weakly stationary X, $X \cap E_r$ is also weakly stationary. Therefore, its intensity measure is determined by the shape distribution $P_0^{E_r}$ and the intensity γ_{E_r}. Our goal is to express γ_{E_r}, the mean values $P_0^{E_r}(V_j)$, and the quermass densities $D_j(X \cap E_r)$ by means of γ, $P_0(V_k)$, and $D_k(X)$. For this purpose, let $K_0 \subset E_r$ be r-dimensional, $r \geq j$, $\beta(Z) = \mathbf{R}^d$, and β_0 in the relative interior of K_0. Then

$$\Psi_j(K_0 \cap Z, \beta_0) = \Psi_j(Z \cap E_r, \beta_0).$$

Also,

$$\varphi_k^{(j)}(K_0, Z, \beta_0 \times \mathbf{R}^d) = 0,$$

for $k \neq r$, and (for $Z = K + L$ and $r - j \leq d - q$)

$$\varphi_r^{(j)}(K_0, Z, \beta_0 \times \mathbf{R}^d)$$

$$= \frac{\binom{d}{q}\binom{d-q}{r-j}}{\kappa_{r-j}\kappa_q} V(\underbrace{K, \ldots, K}_{d+j-r-q}, \underbrace{B_{E_r}, \ldots, B_{E_r}}_{r-j}, \underbrace{B_L, \ldots, B_L}_{q}) \cdot \lambda_{E_r}(\beta_0).$$

This follows for polytopes K, K_0 (in general relative position) from (2.4) and (2.16), for K, $K_0 \in \mathcal{K}_d$ by approximation, and for K, $K_0 \in \mathcal{R}_d$ by additivity. Hence, we conclude from (4.1)

$$\mathbb{E} \sum_{Z \in X} \Psi_j(Z \cap E_r, \cdot)$$

$$= \frac{\binom{d}{q}\binom{d-q}{r-j}}{\kappa_{r-j}\kappa_q} \times$$

$$\times \gamma \left(\int_{\mathcal{Z}_q^0} V(\underbrace{K, \ldots, K}_{d+j-q-r}, \underbrace{B_{E_r}, \ldots, B_{E_r}}_{r-j}, \underbrace{B_L, \ldots, B_L}_{q}) \, dP_0(K + L) \right) \cdot \lambda_{E_r}$$

for $j = p, \ldots, r$.

On the other hand,

$$\mathbb{E} \sum_{Z \in X} \Psi_j(Z \cap E_r, \cdot) = \mathbb{E} \sum_{Z \in X \cap E_r} \Psi_j(Z, \cdot)$$

$$= D_j(X \cap E_r) \cdot \lambda_{E_r}$$

by Theorem 4.2. Thus, we have proved the following formula for the quermass densities of the induced point process $X \cap E_r$.

THEOREM 5.1. *We have*

$$D_j(X \cap E_r)$$

$$= \frac{\binom{d}{q}\binom{d-q}{r-j}}{\kappa_{r-j}\kappa_q} \times$$

$$\times \gamma \int_{\mathscr{X}_q^0} V(\underbrace{K, \ldots, K}_{d+j-q-r}, \underbrace{B_{E_r}, \ldots, B_{E_r}}_{r-j}, \underbrace{B_L, \ldots, B_L}_{q}) \, dP_0(K + L),$$

for $j = p, \ldots, r$.

If the cylinders of the process X are all convex, then $D_p(X \cap E_r) = \gamma_{E_r}$ for all E_r, hence, Theorem 5.1 gives an expression for γ_{E_r},

$$\gamma_{E_r} = \begin{cases} \dfrac{\binom{d}{q}\binom{d-q}{r}}{\kappa_r \kappa_q} \gamma \times \\ \times \displaystyle\int_{\mathscr{X}_q^0} V(\underbrace{K, \ldots, K}_{d-q-r}, \underbrace{B_{E_r}, \ldots, B_{E_r}}_{r}, \underbrace{B_L, \ldots, B_L}_{q}) \, dP_0(K + L), \quad q + r < d, \\[20pt] \gamma \displaystyle\int_{\mathscr{X}_q^0} [E_r, L] \, dP_0(K + L), \quad q + r \geq d. \end{cases}$$

$$(5.1)$$

Here, we have used

$$V(\underbrace{B_{E_r}, \ldots, B_{E_r}}_{d-q}, \underbrace{B_L, \ldots, B_L}_{q}) = \frac{\kappa_{d-q}\kappa_q}{\binom{d}{q}} [E_r, L].$$

Also, in this case, Theorem 3.2 implies

$$P_0^{E_r}(V_j)$$

$$= \frac{\binom{d}{q}\binom{d-q}{r-j}}{\kappa_{r-j}\kappa_q} \frac{\gamma}{\gamma_{E_r}} \times$$

$$\times \int_{\mathscr{X}_q^0} V(\underbrace{K, \ldots, K}_{d+j-q-r}, \underbrace{B_{E_r}, \ldots, B_{E_r}}_{r-j}, \underbrace{B_L, \ldots, B_L}_{q}) \, dP_0(K + L). \quad (5.2)$$

Matheron (1975) has corresponding formulas for Poisson processes of convex particles. In order to show the connection, let X be a process of convex particles. Since

$$V(\underbrace{K, \ldots, K}_{d-r}, \underbrace{B_{E_r}, \ldots, B_{E_r}}_{r})$$

is (up to a constant) the $(d-r)$-content of the projection $K \,|\, E_r^{\perp}$ of K onto $E_r^{\perp}$, we get from (5.1)

$$\gamma_{E_r} = \binom{d}{r} \cdot \gamma P_0(V_{d-r}(\cdot \,|\, E_r^{\perp})). \tag{5.3}$$

For $j \geq 1$, the following projection formula can be deduced, e.g., from (9.7) in Schneider and Weil (1983):

$$\int_{\mathscr{L}_{r-j}^{E_r}} V_{d+j-r}(K \,|\, M^{\perp}) \, d\nu_{r-j}^{E_r}(M)$$

$$= \frac{\binom{d}{r-j} \kappa_j}{\binom{r}{j} \kappa_{d+j-r}} V(\underbrace{K, \ldots, K}_{d+j-r}, \underbrace{B_{E_r}, \ldots, B_{E_r}}_{r-j}). \tag{5.4}$$

Here $\nu_{r-j}^{E_r}$ is the normalized invariant measure on $\mathscr{L}_{r-j}^{E_r}$, the space of $(r-j)$-dimensional subspaces in E_r. Therefore, (5.2) implies

$$P_0^{E_r}(V_j) = \frac{\binom{r}{j} \kappa_{d+j-r}}{\kappa_{r-j} \kappa_j} \frac{\gamma}{\gamma_{E_r}} \int_{\mathscr{L}_{r-j}^{E_r}} P_0(V_{d+j-r}(\cdot \,|\, M^{\perp})) \, d\nu_{r-j}^{E_r}(M), \tag{5.5}$$

$j = 1, \ldots, r$.

Equations (5.3) and (5.5) are Matheron's formulas; as we have seen, they are valid without Poisson assumptions.

If X is weakly isotropic, then we use (2.8) instead of (2.4) and get

$$D_j(X \cap E_r) = \alpha_{djr} \gamma P_0(V_{d+j-r}), \tag{5.6}$$

$j = p, \ldots, r$. For convex cylinders, (5.6) implies

$$\gamma_{E_r} = \begin{cases} \alpha_{d0r} \gamma P_0(V_{d-r}), & q+r < d \\ \alpha_{d,q+r-d,r}, & q+r \geq d, \end{cases} \tag{5.7}$$

and

$$P_0^{E_r}(V_j) = \alpha_{djr} \frac{\gamma}{\gamma_{E_r}} P_0(V_{d+j-r}). \tag{5.8}$$

For Poisson processes of particles $(q=0)$, (5.7) and (5.8) were first proved by

Matheron (1975); his result was extended to non-Poissonian processes by Stoyan (1979), (1982).

If X is a process of flats, Theorem 5.1 implies that

$$D_j(X \cap E_r) = 0,$$

for all j if $q + r < d$. In particular, $\gamma_{E_r} = D_0(X \cap E_r) = 0$, hence $X \cap E_r$ is the 'empty process', as we have mentioned earlier. If $q + r \geq d$, then

$$D_j(X \cap E_r) = 0$$

for $j = q + r - d + 1, \ldots, r$. The following formula for $D_{q+r-d}(X \cap E_r)$ is just the second part of (5.1).

THEOREM 5.2. *Let X be a point process of q-flats with $q + r \geq d$. Then*

$$\gamma_{E_r} = \gamma \int_{\mathscr{L}_q^d} [E_r, L] \, dP_0(L).$$

Equation (5.7) gives the corresponding result for weakly isotropic X. Again, for Poisson processes of flats, similar formulas are obtained in Matheron (1975).

Finally, let us assume that X is a process of j-sets $K \in \mathscr{R}_d$, $j \in \{0, \ldots, d-1\}$. Then we can strengthen the formula in Theorem 5.1. For a j-set $K \in \mathscr{R}_d$, there are sets $K_1, \ldots, K_n \in \mathscr{R}_d$ of dimension at most j with $K = \bigcup_{i=1}^n K_n$ and

$$\Psi_j(K, \cdot) = \sum_{i=1}^n \Psi_j(K_i, \cdot),$$

Let

$$\eta_K = \sum_{i=1}^n V_j(K_i) \, \epsilon_{L(K_i)} \tag{5.9}$$

where $\epsilon_{L(K_i)}$ is the Dirac measure on $L(K_i) \in \mathscr{L}_j^d$, a subspace parallel to the affine hull of K_i. It is easily seen that η_K depends only on K and not on the special representation $K = \bigcup_{i=1}^n K_i$. η_K is a measure on $\mathscr{L}_j^d$ with $\eta_K(\mathscr{L}_j^d) = V_j(K)$, hence

$$\tilde{P}_0 = \frac{1}{P_0(V_j)} P_0(\eta.)$$

is a probability measure on $\mathscr{L}_j^d$. (Here we have to assume $P_0(V_j) > 0$, but the case $P_0(V_j) = 0$ is of no interest for the following. We also leave out the details for the measurability of $K \mapsto \eta_K$, which follows from the definition of X by counting measures.)

We call $\tilde{P}_0$ the *directional distribution* of X.

THEOREM 5.3. *For a process X of j-sets $K \in \mathscr{R}_d$ we have*

$$D_{j+r-d}(X \cap E_r)$$

$$= D_j(X) \int_{\mathscr{L}_j^d} [L, E_r] \, d\tilde{P}_0(L), \quad r = d - j, \ldots, d.$$

Proof. From Theorem 5.1, we know that

$$D_{j+r-d}(X \cap E_r)$$

$$= \frac{\binom{d}{j}}{\kappa_{d-j}} \gamma \int_{\mathfrak{R}_d^0} V(\underbrace{K, \ldots, K}_{j}, \underbrace{B_{E_r}, \ldots, B_{E_r}}_{d-j}) \, dP_0(K).$$

Using the representation $K = \bigcup_{i=1}^{n} K_i$ underlying (5.9), we get

$$V(\underbrace{K, \ldots, K}_{j}, \underbrace{B_{E_r}, \ldots, B_{E_r}}_{d-j})$$

$$= \sum_{i=1}^{n} V(\underbrace{K_i, \ldots, K_i}_{j}, \underbrace{B_{E_r}, \ldots, B_{E_r}}_{d-j})$$

$$= \frac{\kappa_{d-j}}{\binom{d}{j}} \sum_{i=1}^{n} V_j(K_i)[L(K_i), E_r]$$

$$= \frac{\kappa_{d-j}}{\binom{d}{j}} \int_{\mathscr{L}_j^d} [L, E_r] \, d\eta_K(L),$$

hence

$$D_{j+r-d}(X \cap E_r)$$

$$= \gamma \int_{\mathscr{L}_j^d} [L, E_r] \, d(P_0(\eta.)(L))$$

$$= D_j(X) \int_{\mathscr{L}_j^d} [L, E_r] \, d\tilde{P}_0(L).$$

For a process X of j-sets we may consider the union set Y,

$$Y(\omega) = \bigcup_{K \in X(\omega)} K,$$

which is a weakly stationary, random closed set in $\mathbf{R}^d$. If the particles of X do not overlap (i.e., if their intersections are $(j-1)$-sets), then the quermass densities $D_j(X)$ and $D_{j+r-d}(X \cap E_r)$ depend only on Y.

More precisely, from Theorem 4.2 it follows that $D_j(X) = D_j(Y)$ and

$$D_{j+r-d}(X \cap E_r) = D_{j+r-d}(Y \cap E_r)$$

(see Weil (1983b) for details on quermass densities of random sets). Also, the directional distribution $\tilde{P}_0$ of X depends only on Y. To see this, the following formula can be derived similarly to (4.3). Let $\mathscr{A}$ be a Borel set in $\mathscr{L}_j^d$ and, for a j-set $K \in \mathfrak{R}_d$, let $\beta(K, \mathscr{A})$ be the closure of all midpoints of j-dimensional balls in K which are parallel to subspaces in $\mathscr{A}$. As one can easily show, $K \mapsto \beta(K, \mathscr{A})$ is measurable and translation covariant. Let K_0 be a ball with $V_d(K_0) = 1$. Then

$$\tilde{P}_0(\mathscr{A}) = \frac{1}{D_j(Y)} E\Psi_j(K_0 \cap Y, \beta(K_0 \cap Y, \mathscr{A})).$$

For stationary fibre processes $(j = 1)$ in $\mathbf{R}^2$ or $\mathbf{R}^3$ and for stationary surface processes $(j = 2)$ in $\mathbf{R}^3$, Theorem 5.3 was obtained by Mecke and Stoyan (1980a), Mecke and Nagel (1980), and Pohlmann, Mecke, and Stoyan (1981) (see also Stoyan, Mecke and Pohlmann (1980) for a special nonisotropic fibre process in $\mathbf{R}^2$). Here fibre processes and surface processes are point processes on more general set classes but, as we have mentioned in the introduction, our model can serve as an approximation. As a more general result, which contains Theorem 5.2 and Theorem 5.3 as special cases, Zähle (1982) proved intersection formulas for point processes of Hausdorff rectifiable closed sets. The correspondence of $\tilde{P}_0$ with the distribution of tangents or normals in the above-mentioned papers can be seen from (5.9). We finally remark that $\tilde{P}_0$ can be interpreted as the distribution of the tangent space in a 'typical point' of the random set Y.

In order to obtain a second point process in E_r we consider the case of projected thick sections. Let the convex cylinder $Z_0 = K_0 + E_r$ be a thickening of the flat E_r by a set $K_0 \in \mathscr{K}_d$, $K_0 \subset E_r^\perp$. If X is a weakly stationary point process of convex particles, we denote by $(X \cap Z_0) | E_r$ the collection of projected intersections $(K \cap Z_0) | E_r$, $K \in X$. $(X \cap Z_0) | E_r$ is also a weakly stationary process of convex particles (in E_r). We denote its intensity by γ_{Z_0}, and its shape distribution by $P_0^{Z_0}$. In order to get formulas for γ_{Z_0}, $P_0^{Z_0}(V_j)$, and $D_j((X \cap Z_0) | E_r)$, let β be a Borel set in the relative interior of a convex body $K' \subset E_r$. As in Section 4, we get from the Campbell theorem (3.2)

$$E \sum_{K \in X} \Psi_j((K \cap Z_0) | E_r, \beta)$$

$$= \gamma \int_{\mathscr{K}_d^0} \int_{\mathbf{R}^d} \Psi_j(((K + x) \cap Z_0) | E_r, \beta) \, d\lambda_d(x) \, dP_0(K) \qquad (5.10)$$

$$= \gamma \int_{\mathscr{K}_d^0} \int_{E_r} \int_{E_r^\perp} \Psi_j(((K + x + y) \cap Z_0) | E_r, \beta) \, d\lambda_{E_r^\perp}(y) \, d\lambda_{E_r}(x) \, dP_0(K).$$

Using (2.11), we can simplify the inner integral

$$\int_{E_r^\perp} \Psi_j(((K + x + y) \cap Z_0) | E_r, \beta) \, d\lambda_{E_r^\perp}(y)$$

$$= \int_{E_r^\perp} \Psi_j(((K + x) \cap (Z_0 + y)) | E_r, \beta) \, d\lambda_{E_r^\perp}(y) \qquad (5.11)$$

$$= \varphi_{d+j-r}^{(j)}(K + (-K_0) + x, E_r, \beta + E_r^\perp).$$

If K, K_0 are polytopes in general relative position, (2.4) implies

$$\int_{E_r} \varphi_{d+j-r}^{(j)}(K+(-K_0)+x, E_r, \beta+E_r^\perp)\, d\lambda_{E_r}(x)$$

$$= \int_{E_r} \sum_{F\in\mathcal{F}_{d+j-r}(K+(-K_0)+x)} \gamma(F, E_r, K+(-K_0)+x, E_r)\times$$

$$\times [L(F), E_r]\lambda_F(\beta+E_r^\perp)\, d\lambda_{E_r}(x) \tag{5.12}$$

$$= \sum_{F\in\mathcal{F}_{d+j-r}(K+(-K_0))} \gamma(F, E_r, K+(-K_0), E_r)[L(F), E_r]\times$$

$$\times \left(\int_{E_r} \lambda_F(\beta+x+E_r^\perp)\, d\lambda_{E_r}(x)\right).$$

By (2.3)

$$\int_{E_r} \lambda_F(\beta+x+E_r^\perp)\, d\lambda_{E_r}(x)$$

$$= \int_{E_r} \Psi_{d+j-r}(F\cap(K'+E_r^\perp+x), \beta+E_r^\perp+x)\, d\lambda_{E_r}(x) \tag{5.13}$$

$$= V_{d+j-r}(F)\lambda_{E_r}(\beta).$$

Combining (5.12) and (5.13) and using (2.13) and the multilinearity of mixed volumes, we get

$$\int_{E_r} \varphi_{d+j-r}^{(j)}(K+(-K_0)+x, E_r, \beta+E_r^\perp)\, d\lambda_{E_r}(x)$$

$$= \Phi_{d+j-r}^{(j)}(K+(-K_0), E_r)\lambda_{E_r}(\beta)$$

$$= \frac{\binom{d}{r-j}}{\kappa_{r-j}} \sum_{i=j}^{d+j-r} \binom{d+j-r}{i} V(\underbrace{K,\ldots,K}_{i}, \underbrace{-K_0,\ldots,-K_0}_{d+j-r-i}, \underbrace{B_{E_r},\ldots,B_{E_r}}_{r-j}) \cdot \lambda_{E_r}(\beta),$$

first for polytopes in general relative position, then for arbitrary K, K_0 by approximation.

In view of (5.10), (5.11), and Theorem 4.2, this result gives a formula for $D_j((X\cap Z_0)\mid E_r)$.

THEOREM 5.4. *We have*

$$D_j((X\cap Z_0)\mid E_r)$$

$$= \frac{\binom{d}{r-j}}{\kappa_{r-j}} \gamma \sum_{i=j}^{d+j-r} \binom{d+j-r}{i} \times$$

$$\times \int_{\mathcal{K}_d^0} V(\underbrace{K,\ldots,K}_{i}, \underbrace{-K_0,\ldots,-K_0}_{d+j-r-i}, \underbrace{B_{E_r},\ldots,B_{E_r}}_{r-j})\, dP_0(K),$$

for $j=0,\ldots,r$.

For weakly isotropic X,

$$\int_{\mathcal{H}_d^0} V(\underbrace{K,\ldots,K}_{i}, \underbrace{-K_0,\ldots,-K_0}_{d+j-r-i}, \underbrace{B_{E_r},\ldots,B_{E_r}}_{r-j})\, dP_0(K)$$

$$= \int_{SO_d} \int_{\mathcal{H}_d^0} V(\underbrace{\theta K,\ldots,\theta K}_{i}, \underbrace{-K_0,\ldots,-K_0}_{d+j-r-i}, \underbrace{B_{E_r},\ldots,B_{E_r}}_{r-j})\, dP_0(K)\, d\nu(\theta)$$

$$= \int_{\mathcal{H}_d^0} \int_{SO_d} V(\underbrace{K,\ldots,K}_{i}, \underbrace{-\theta K_0,\ldots,-\theta K_0}_{d+j-r-i}, \underbrace{\theta B_{E_r},\ldots,\theta B_{E_r}}_{r-j})\, d\nu(\theta)\, dP_0(K)$$

$$= \frac{\kappa_{r-j}}{\binom{d}{r-j}} \gamma_{djir} P_0(V_i) V_{d+j-r-i}(K_0),$$

where we have used a rotation formula in Schneider (1981). Thus, Theorem 5.4 implies

$$D_j((X\cap Z_0)\,|\,E_r) = \gamma \sum_{i=j}^{d+j-r} \gamma_{djir} P_0(V_i) V_{d+j-r-i}(K_0), \qquad (5.14)$$

$j = 0, \ldots, r$.

Since $\gamma_{Z_0} = D_0((X\cap Z_0)\,|\,E_r)$, formulas for γ_{Z_0} and $P_0^{Z_0}(V_j)$ follow from Theorem 5.4 and (5.14). For circular cylinders Z_0, (5.14) is due to Davy (1976). As we have mentioned, (2.11) and (2.12) can be generalized to sets in $\mathcal{R}_d$ with appropriate modifications. We only consider the case of a fibre process X, i.e., a process of 1-sets. Here

$$D_1((X\cap Z_0)\,|\,E_r)$$

$$= \frac{r\binom{d}{r}}{\kappa_{r-1}} \gamma \int_{\mathcal{R}_d^0} V(K, \underbrace{-K_0,\ldots,-K_0}_{d-r}, \underbrace{B_{E_r},\ldots,B_{E_r}}_{r-1})\, dP_0(K),$$

which can be simplified using the directional distribution $\tilde{P}_0$

$$D_1((X\cap Z_0)\,|\,E_r)$$

$$= \frac{r\binom{d}{r}}{\kappa_1 \kappa_{r-1}} D_1(X) \int_{\mathcal{L}_1^d} V(B_L, \underbrace{K_0,\ldots,K_0}_{d-r}, \underbrace{B_{E_r},\ldots,B_{E_r}}_{r-1})\, d\tilde{P}_0(L). \qquad (5.15)$$

If X is weakly isotropic, then

$$D_1((X\cap Z_0)\,|\,E_r) = \frac{r\kappa_r \kappa_{d-1}}{d\kappa_{r-1}\kappa_d} D_1(X) V_{d-r}(K_0). \qquad (5.16)$$

If the fibres of the process X have no segments in common, i.e., if the intersections of different fibres are 0-sets, then for μ_r-almost all E_r and for $d \geq 3$, $D_1((X\cap Z_0)\,|\,E_r)$ depends only on the union set of $(X\cap Z_0)\,|\,E_r$.

For $D_0((X\cap Z_0)\,|\,E_r)$ or γ_{Z_0} no simple formulas exist, in general. If X is a weakly stationary process of simply connected fibres each fibre $K \in X$ intersects Z_0 in finitely many (simply) connected parts. In order to determine γ_{Z_0}, not only the number of these parts must be recognized from their projections onto E_r but also it must be clear which parts in Z_0 come from the same original fibre $K \in X$. Quite often this is impossible and it is much more realistic to consider the different connected parts of $K \cap Z_0$, $K \in X$, as the fibres of a new fibre process (in Z_0) which are then projected onto E_r. Let the resulting fibre process in E_r be denoted by $\hat{X}_{Z_0}$ and its intensity by $\tilde{\gamma}_{Z_0}$. For $\tilde{\gamma}_{Z_0}$, a simple formula can be given. By construction of $\hat{X}_{Z_0}$ and from Theorem 3.1 we have

$$\tilde{\gamma}_{Z_0} = \lim_{t \to \infty} \frac{1}{V_r(tB_{E_r})} \mathbb{E} \sum_{K \in X} V_0(K \cap (K_0 + tB_{E_r})).$$

Now we can use (4.8)

$$\tilde{\gamma}_{Z_0} = \lim_{t \to \infty} \frac{1}{V_r(tB_{E_r})} \gamma [V_d(K_0 + tB_{E_r}) +$$
$$+ dP_0(V(-K_0 + tB_{E_r}, \ldots, -K_0 + tB_{E_r}, \cdot))]$$
$$= \lim_{t \to \infty} \frac{1}{V_r(tB_{E_r})} \gamma [V_{d-r}(K_0) V_r(tB_{E_r}) +$$
$$+ dP_0(V(\underbrace{-K_0, \ldots, -K_0}_{d-r}, \underbrace{tB_{E_r}, \ldots, tB_{E_r}}_{r-1}, \cdot)) +$$
$$+ dP_0(V(\underbrace{-K_0, \ldots, -K_0}_{d-r-1}, \underbrace{tB_{E_r}, \ldots, tB_{E_r}}_{r}, \cdot))],$$

hence

$$\tilde{\gamma}_{Z_0} = \gamma \left[V_{d-r}(K_0) + \frac{d}{\kappa_r} P_0(V(\underbrace{K_0, \ldots, K_0}_{d-r-1}, \underbrace{B_{E_r}, \ldots, B_{E_r}}_{r}, \cdot)) \right]. \tag{5.17}$$

Analogously to (5.15), a simplification of (5.17) is possible using $\tilde{P}_0$. For weakly isotropic X, (5.17) becomes

$$\tilde{\gamma}_{Z_0} = \gamma \left[V_{d-r}(K_0) + \frac{2\kappa_{d-1}}{d\kappa_d} V_{d-r-1}(K_0) P_0(V_1) \right]. \tag{5.18}$$

For $d = 3$ and $r = 2$, (5.15) and (5.16) were proved by Nagel (1983); for arbitrary d and $r = d - 1$, (5.16) and (5.18) are due to Zähle (1984).

6. Discussion and Comments

Before we study a few applications of the results obtained so far, we want to discuss some generalizations and interrelationships between the formulas. Also, some comments on the literature are in order here.

We have studied point processes X of cylinders not only because they are interesting models for several natural phenomena, but also as a unifying notion for both, point processes of particles and point processes of flats. An important step in our considerations was the decomposition (3.1) of the intensity measure, which was based on the isomorphism i. i^{-1} transforms X into a point process $\tilde{X}$ on $\tilde{\mathcal{L}}_q \subset \mathbf{R}^d \times \mathcal{L}_q^0$. $\tilde{X}$ can be interpreted as a marked point process in $\mathbf{R}^d$, given by an underlying point process $\bar{X}$ in $\mathbf{R}^d$ (which is the process of centers $s(K)$, $K + L \in X$) and with mark space $\mathcal{L}_q^0$. Some authors prefer to work with these marked point processes instead of X. For processes of particles, the two concepts are indeed equivalent (as long as the mark space contains only the shapes $\mathcal{K}_d^0$ and no additional information). For cylinder processes our approach is preferable since then the mappings $X \to \tilde{X}$ and $X \to \bar{X}$ do not preserve invariance properties, i.e., for stationary X, $\tilde{X}$ and $\bar{X}$ need not be stationary. For a stationary process of particles, γ is the intensity of the underlying ordinary (and stationary) point process $\bar{X}$ and P_0 is the mark distribution. Of course, $\bar{X}$ may have multiple points even if X is simple.

As an even more general notion, point processes X of closed sets may be considered (see Matheron (1975)). Especially, if the sets $C \in X$ are in the class $\mathcal{S}_d$ of locally finite, countable unions of convex bodies, then the curvature measures $\Psi_j(C, \cdot)$, $j = 0, \ldots, d$, exist as (signed) Radon measures. However, a decomposition of the type (3.1) is more difficult in this case and, therefore, methods and results of the last sections do not immediately generalize to such processes X. For weakly stationary X, the quermass densities $D_j(X)$ can be defined by

$$\mathsf{E} \sum_{C \in X} \Psi_j(C, \cdot) = D_j(X) \cdot \lambda_d, j = 0, \ldots, d. \tag{6.1}$$

Here, of course, some conditions on X are necessary (similar to the conditions in Theorem 3.2) which guarantee that $\mathsf{E} \sum_{C \in X} \Psi_j(C, \cdot)$ is a locally finite (signed) Radon measure on $\mathbf{R}^d$. (6.1) then follows from the stationarity of Θ and the Campbell theorem. If X is, moreover, weakly isotropic, a modification of the argument in Section 3 (see Zähle (1986) for a similar proof) leads to the following analog of (4.7)

$$\mathsf{E} \sum_{C \in X} \Psi_j(K_0 \cap C, \beta_0) = \sum_{k=j}^{d} \alpha_{djk} \Psi_k(K_0, \beta_0) \cdot D_{d+j-k}(X), \tag{6.2}$$

$j = 0, \ldots, d.$

In the same way, (5.6) and (5.14) can be generalized to processes X on $\mathcal{S}_d$:

$$D_j(X \cap E_r) = \alpha_{djr} D_{d+j-r}(X), \qquad\qquad j = 0, \ldots, r, \tag{6.3}$$

$$D_j((X \cap Z_0) \mid E_r) = \sum_{i=j}^{d+j-r} \gamma_{djir} D_i(X) V_{d+j-r-i}(K_0), \qquad j = 0, \ldots, r. \tag{6.4}$$

Since cylinder processes X are special processes on $\mathcal{S}_d$, (6.4) implies that (5.14) is true for processes of cylinders, too.

There is an obvious connection between point processes on $\mathcal{L}_d$ (or $\mathcal{S}_d$) and random $\mathcal{S}_d$-sets, since for each point process X we can consider its union set Y. If the curvature measures obey

$$\Psi_j(Y, \cdot) = \sum_{C \in X} \Psi_j(C, \cdot)$$

(this is the case if the sets $C \in X$ have at most $(j-1)$-dimensional intersections), then

$$D_j(Y) = D_j(X),$$

hence, we have similar formulas for the jth quermass density of X and Y. Such formulas for random $\mathcal{S}_d$-sets are obtained in Davy (1978), Weil (1983a), (1984), Weil and Wieacker (1984). For more general random sets, see Mecke (1981a) and Zähle (1982), (1986). A result in the opposite direction was obtained by Weil and Wieacker (1984) who show that any random $\mathcal{S}_d$-set Y is the union set of some point process X on $\mathcal{K}_d$. The construction can be modified so that X has the same invariance properties as Y (Weil and Wieacker (1986)). The connection between X and Y has led some authors to use the term 'process' in some cases for Y, too. For instance, fibre processes are sometimes random aggregates of fibres, where the individual fibres cannot be distinguished.

For a process X of q-flats, the intensity γ coincides with $D_q(X)$, the mean q-content of X per unit volume (Theorem 4.3). Thus, the intensity γ depends only on the union set Y of X. For a process X of q-sets (without overlapping), the intensity γ and $D_q(X)$ are different quantities. While γ depends strongly on the number of different particles of X, $D_q(X)$ is the same as $D_q(Y)$ (where Y is again the union set). For this reason, some authors call $D_q(X)$ the intensity of X, especially if the process is only given by its union set Y.

From the intersection formulas which we have given, it is quite easy to obtain formulas for quermass densities of intersections and superpositions of two (or more) independent processes. The invariance conditions must be imposed only on one of the processes. In particular, this means that there are generalizations of the formulas of Section 5, where the r-flat E_r is replaced by an r-set K_0 or a (not necessarily stationary) point process X_0 of r-sets. Formulas of this kind have been investigated in Mecke (1981b) and Zähle (1986).

As we mentioned, the quermass densities of the process X and the union set Y are the same only in special cases. General formulas for the connection between $D_j(X)$ and $D_j(Y)$ are known only for Poisson processes (see Davy, 1976, 1978; Wieacker, 1982; Weil, 1983a; Kellerer, 1984; Weil and Wieacker, 1984; Zähle, 1986 for more details).

We have exploited the basic formula (3.2) only for the curvature measures since we aimed to apply the integral formulas of Section 2. It is obvious that the procedure underlying Section 4 can be performed if the curvature measures are multiplied by a 'weighting factor' $g = g(Z)$ which depends on $Z \in \mathcal{Z}_q$ in a translation invariant way. For example, (4.1) then reads

$$\mathbb{E} \sum_{Z \in X} g(Z)\, \Psi_j(K_0 \cap Z, \beta_0 \cap \beta(Z))$$

$$= \gamma \bigg[\Psi_j(K_0, \beta_0) \int_{\mathscr{L}_q^0} g(K + L)\Psi_{d-q}(K, \beta(K))\, dP_0(K + L) +$$

$$+ \sum_{k=j+1}^{d-1} \int_{\mathscr{L}_q^0} g(Z)\varphi_k^{(j)}(K_0, Z, \beta_0 \times \beta(Z))\, dP_0(Z) + \tag{6.5}$$

$$+ \Psi_d(K_0, \beta_0) \int_{\mathscr{L}_q^0} g(K + L)\, \Psi_{j-q}(K, \beta(K))\, dP_0(K + L) \bigg],$$

$$j = 0, \ldots, d.$$

Analogously, the other formulas in Sections 3 and 4 can be modified. Of course, the integrability conditions on X must be changed according to g, too. The variants which follow from this concept are too numerous to mention them all. We will use some special weighting factors in the following section. We give only two further results of this kind. If X is a process of convex cylinders, $K_0 \in \mathscr{K}_d$, and $g = V_j$, then using (4.2) we have

$$\mathbb{E} \sum_{Z \in X} V_j(Z) 1_{\mathscr{L}_q(K_0)}(Z)$$

$$= \gamma \bigg[P_0(V_j \cdot V_d) + \sum_{k=1}^{d-q} P_0(\Phi_k^{(0)}(K_0, \cdot) \cdot V_j) \bigg]. \tag{6.6}$$

If K_0 is a point x_0, (6.6) holds for arbitrary cylinder processes X and becomes

$$\mathbb{E} \sum_{\substack{Z \in X, \\ x_0 \in Z}} V_j(Z) = \gamma P_0(V_j \cdot V_d).$$

If X is a process of q-flats, $E_r \in \mathscr{E}_r^d$, $\mathscr{B} \subset \mathscr{L}_q^d$ a Borel set, $\tilde{\mathscr{B}} = \{L + x \mid L \in \mathscr{B}, x \in L^\perp\}$, and

$$g(L) = \begin{cases} [L, E_r]^{-1} & \text{if } L \in \tilde{\mathscr{B}} \\ 0 & \text{if } L \notin \tilde{\mathscr{B}}, \end{cases}$$

then

$$\mathbb{E} \sum_{\substack{L \in X, \\ L \in \tilde{\mathscr{B}}}} \frac{1}{[L, E_r]} \Psi_{q+r-d}(L \cap E_r, \cdot)$$

$$= \gamma P_0(\mathscr{B})\lambda_{E_r}, \tag{6.7}$$

by Theorem 5.2 and the argument before Theorem 5.1. Similarly, for a process X of convex q-sets,

$$\mathbb{E} \sum_{\substack{K \in X \\ L(K) \in \mathscr{B}}} \frac{1}{[L(K), E_r]} \Psi_{q+r-d}(K \cap E_r, \cdot)$$

$$= D_q(X) \tilde{P}_0(\mathscr{B})\lambda_{E_r}. \tag{6.8}$$

Since each process X of q-sets can be decomposed into a process X' of convex q-sets with the same density $D_q(X) = D_q(X')$, (6.8) holds for arbitrary (weakly stationary) processes of q-sets with the appropriate definition of $L(K)$. Special mention should be made of the case $q = d - 1$. Here, the result applies to the process of boundaries of a particle process X, provided bd $K \in \mathcal{R}_d$ for $K \in X$ (e.g., boundaries of convex polytope processes). For dimensions 2 and 3, (6.7) and (6.8) are due to Ambartzumian (1977), (1982), Mecke and Stoyan (1980a), Pohlmann, Mecke and Stoyan (1981). Again, there are similar formulas with E_r replaced by an r-set K_0 or a process X_0 of r-sets (e.g., Ohser (1981)).

Formula (6.5) and its consequences have a close connection to another famous problem in stereology. If the shape distribution P_0 is concentrated on a class $\tilde{\mathcal{X}} \subset \mathcal{X}_q^0$, the elements of which are characterized by one real parameter $h(Z)$ (e.g., diameter of balls), or if we are interested only in one real parameter $h(Z)$ for $Z \in \mathcal{X}_q^0$ and its distribution P_h (which is the image of P_0 under h), then we may choose h as weight g in (6.5). Consequently, we obtain results analogous to those of Section 5 in which the distribution of h for those $Z \in X$ with $Z \cap E_r \neq \emptyset$ is related to P_h. Notice that these formulas involve $h(Z)$, for $Z \cap E_r \neq \emptyset$, $Z \in X$, and not $h(Z \cap E_r)$. This is a variant of the classical Wicksell problem; formulas for processes of balls, discs, etc. are investigated by Mecke and Stoyan (1980), Pohlmann, Mecke and Stoyan (1981).

Finally, we mention a special class of processes X of particles, the random mosaics. X is a random mosaic, if the union set of X is almost surely $\mathbf{R}^d$ and if the intersection $K \cap K'$ of different particles $K, K' \in X(\omega)$ is a $(d-1)$-set (for almost all realizations of X). The particles then constitute the cells of the mosaic (random mosaics with cylindrical cells can be defined in a similar way, but are of less interest). The classical case are mosaics with convex cells (i.e., necessarily convex polytopes as cells), random curved mosaics have been treated by Weiss and Zähle (1986). The j-dimensional faces of the cells of a random mosaic X with convex cells form another point process $X^{(j)}$, $j = 0, \ldots, d-1$, with the same invariance properties as X. The main interest is in relations between the quermass densities of the processes $X^{(0)}, \ldots, X^{(d-1)}$ and X. Formulas of this kind are collected in Ambartzumian (1974), Cowan (1980), Mecke (1980), (1984), Radecke (1980), Weiss and Zähle (1986).

7. Applications

The formulas which we have presented deal with the quantities γ and $D_j(X)$, $j = 0, \ldots, d$, and the distribution P_0 (resp. $\tilde{P}_0$) of a given geometrical point process X, as well as corresponding notions of transformed images of X. Therefore, they can be used for the estimation of γ, $D_j(X)$, P_0; in particular, they show which natural estimators are unbiased and which are not. In principal, each of the formulas can be exploited in this way. In the following, we will give some examples which are of special practical interest.

As a first example we consider a typical problem in image analysis related to edge effects in sampling windows. Let X be a stationary process of simply connected particles in $\mathcal{R}_2$. The process X is observed in a sampling window $K_0 \in \mathcal{R}_2$ (with area 1). The mean number of particles γ is to be estimated. Natural estimators of γ are $g_1 =$ 'number of particles intersecting K_0' and $g_2 =$ 'number of particles $K \in X$ with $K \subset K_0$'. Of course, g_1 is an overestimation of γ, and g_2 an underestimation of γ. More precisely, if X is isotropic and if the particles of X and K_0 are convex the bias of g_1 is given by

$$\gamma[2P_0(V_1) \cdot V_1(K_0) + P_0(V_2)]$$

(in view of (4.8)). If in g_1 each particle K is counted with weighting factor

$$[1 + 2V_1(K_0)V_1(K) + V_2(K)]^{-1},$$

the new estimator $\tilde{g}_1$ is unbiased. Analogously, g_2 can be changed into an estimator $\tilde{g}_2$ if each particle $K \in X$, $K \subset K_0$, is weighted by a factor which depends in a more complicated way on the geometry of K and K_0 (see Miles (1974) and Weil (1982) for details). If all particles are small enough w.r.t. K_0, $\tilde{g}_2$ is unbiased. The disadvantages of these estimators are the following. $\tilde{g}_2$ is no longer unbiased if there are particles $K \in X$ which do not fit into K_0. Moreover, if the particles of X and K_0 do not have simple shapes, the calculation of the weighting factor is quite complicated. The weighting factor in $\tilde{g}_1$ can only be determined if $V_1(K)$ and $V_2(K)$, $K \in X$, are known. This requires either knowledge about the shapes in X or the possibility to observe the part $K \setminus K_0$ of the particles $K \in X$ with $K \cap K_0 \neq \emptyset$.

Since γ is also the intensity of the underlying point process X of centers $s(K)$, $K \in X$, an obviously unbiased estimator of γ is $g_3 =$ 'number of particles $K \in X$ with $s(K) \in K_0$'. This method of 'associated points' can be generalized by associating more than one point with each particle and by counting particles $K \in X$ with weights α, $0 \leq \alpha \leq 1$, according to the number of associated points of K which are in K_0. For more details, see Jensen and Sundberg (1985), (1986). A variant of the method of associated points is the tangent count of DeHoff (1978). If the particles do not have simple shapes, the associated points method usually also has to use information outside the sampling window. Another disadvantage is that not all particles which are observed in K_0 are represented in the estimator.

An unbiased estimator which uses the full information of $X \cap K_0$ can be based on the isotropic version of the system of linear equations (4.8), for $j = 0, 1, 2$. Solving this system yields unbiased estimators of $D_0(X)$, $D_1(X)$, $D_2(X)$ which are linear combinations of

$$\sum_{K \in X} V_0(K_0 \cap K), \; \sum_{K \in X} V_1(K_0 \cap K) \; \text{ and } \; \sum_{K \in X} V_2(K_0 \cap K).$$

Of course, here one has to assume that either X is isotropic or K_0 is circular.

This estimator was proposed in the book of Santaló (1976), pp. 282–286; it was studied in more detail by Schwandtke, Ohser and Stoyan (1986). Another estimator of γ which uses the full information is the sum of the total Gaussian curvatures in int K_0,

$$g_4 = \sum_{K \in X} \Psi_0(K, \text{int } K_0).$$

In view of (4.4), g_4 is unbiased. This estimator works for nonisotropic X and for arbitrary shapes $K \in X$ without requiring that the different parts of K in K_0 are recognizable. Since in the plane $\Psi_0(K, \text{int } K_0)$ is determined by the angles between the normals of bd K in the points of bd $K \cap$ bd K_0, the estimator is also mathematically simple. If normals are difficult to determine in practice, DeHoff's tangent count with different directions of tangents can be used as a discrete approximation. Finally, if $K_0 = C_0$ a simple unbiased estimator for γ is given by

$$\sum_{K \in X} [V_0(C_0 \cap K) - V_0(\delta^+ C_0 \cap K)]$$

(see the discussion after Theorem 4.2). For more details about edge effects in particle counting, see the survey of Gundersen (1978) and the recent paper of Schwandtke, Ohser and Stoyan (1986).

Concerning the variances of these estimators, simulations of special point process models with simple shapes have been performed recently by Kellerer (1985), Jensen and Sundberg (1986), Schwandtke, Ohser and Stoyan (1986).

As a second example, we remark that (5.6) contains, for $d = 2$ or 3 and $r = 0$, 1, 2, the set of 'fundamental formulas of stereology'. For simplicity, we consider only the case of a process X of simply connected particles. In the notation which is used in the stereological literature, the volume density $D_d(X)$ is written as V_V, A_A, L_L, P_P, for dimensions $d = 3, 2, 1, 0$, respectively. The surface area density $2D_{d-1}(X)$ is written as S_V, L_A, P_L, according to the dimensions 3, 2, and 1. Moreover, $(2\pi/(d-1)) D_{d-2}(X)$ which is the density of the integral mean curvature is denoted by K_V if $d = 3$, and the number density (or intensity) $D_0(X) = \gamma$ is denoted by N_V and N_A, for $d = 3$ and 2. The same symbols are used for the intersection processes $X \cap E_r$, $r = 0, 1, \ldots, d-1$. The set of fundamental formulas then reads

$$V_V = A_A = L_L = P_P, \quad S_V = \frac{4}{\pi} L_A = 2P_L, \quad K_V = 2\pi N_A. \tag{7.1}$$

For lower dimensional particles (sheets or fibres) the same symbols are used, partially with a different (but obvious) meaning; e.g., for fibre processes in $\mathbb{R}^2$, L_A stands for $D_1(X)$ (and not for $2D_1(X)$). Also other symbols with obvious interpretations may be used (A_V, L_V, P_A, etc.), see Stoyan and Mecke (1983) for the modifications. If the point process X is not weakly isotropic, the formulas

(7.1) still hold for expectations if the sectioning plane is chosen with a random direction (independently of X).

If the intersection processes $X \cap E_r$ are used to get some information about the anisotropy of X then of course one has to work with flats E_r of different nonrandom directions. In particular, this is the case if information about the directional distribution $\tilde{P}_0$ of fields of fibres or sheets is wanted. Since the direction of a plane in $\mathbf{R}^3$ as well as the direction of lines in $\mathbf{R}^2$ or $\mathbf{R}^3$ is determined by a unit vector, the directional distribution $\tilde{P}_0$ for $d = 3$ or 2 can be viewed as an even probability measure on the unit sphere Ω in $\mathbf{R}^3$ or $\mathbf{R}^2$. The quermass densities $D_{j+r-d}(X \cap E_r)$, for $d = 3$ and $r = 1, 2$, or $d = 2$ and $r = 1$, when E_r is supposed to be variable, are functions on $\mathscr{L}_r^d$, hence, they can also be represented as functions on Ω. With the notation used earlier, Theorem 5.3 then gives the formulas

$$L_A(x) = S_V \int_\Omega |\sin \alpha(u, x)| \, d\tilde{P}_0(u) \tag{7.2}$$

and

$$P_L(x) = S_V \int_\Omega |\cos \alpha(u, x)| \, d\tilde{P}_0(u) \tag{7.3}$$

for a process X of sheets in $\mathbf{R}^3$,

$$N_A(x) = L_V \int_\Omega |\cos \alpha(u, x)| \, d\tilde{P}_0(u) \tag{7.4}$$

for a process X of fibres in $\mathbf{R}^3$, and

$$P_L(x) = L_A \int_\Omega |\sin \alpha(u, x)| \, d\tilde{P}_0(u) \tag{7.5}$$

for a process X of fibres in $\mathbf{R}^2$. Here, $\alpha(u, x)$ is the angle between u and x and x is the direction in which the intersection is taken. By a well-known uniqueness result (see the survey of Schneider and Weil (1983)), $\tilde{P}_0$ is uniquely determined by any of the functions on the left sides of (7.2) up to (7.5). The inversion is simple in the case of (7.5), since here $P_L(\cdot)$ is the support function of a centrally symmetric convex body $K_0 \subset \mathbf{R}^2$ and $\tilde{P}_0$ is just the normalized 'length measure' of K_0 (the image of $\Psi_1(K_0, \cdot)$ under the spherical image map from bd K_0 onto Ω). For smooth K_0, $\tilde{P}_0$ can be determined analytically; for a polygon K_0, $\tilde{P}_0$ can be given directly. This also indicates a simple procedure for the estimation of $\tilde{P}_0$ if $P_L(x)$ is only known for finitely many directions x. The inversion of (7.2), (7.3), and (7.4) can also be done analytically (under appropriate smoothness assumptions) but the estimation of $\tilde{P}_0$ from finitely many directions is a much more complicated problem. As we already mentioned in the last section, the situation is much easier if it is possible to observe in the plane (line) of intersection E_r the angles of the sheets or fibres with E_r. Then, $\tilde{P}_0$ can be estimated directly from

(6.8). Another method is possible if the fibres are replaced by cylindrical tubes (Stoyan (1984), (1985b)).

Finally, we want to discuss the situation of projected thick sections. The most interesting case for applications is that of a circular cylinder $Z_0 = E_1 + tB_{E_1^\perp}$ or a thickened plane $Z_0 = E_2 + tB_{E_2^\perp}$ in $\mathbf{R}^3$, $t > 0$.

If we first consider a weakly isotropic field of convex particles, then the quermass density on the left side of (5.14) is a function of t alone and we get for the two situations mentioned above

$$P_L(t) = \tfrac{1}{2} S_V + t K_V + 2\pi t^2 N_V,$$

$$L_L(t) = V_V + \frac{\pi}{4} t S_V + \tfrac{1}{2} t^2 K_V, \tag{7.6}$$

and

$$N_A(t) = \frac{1}{2\pi} K_V + 2t N_V,$$

$$L_A(t) = \frac{\pi}{4} S_V + t K_V, \tag{7.7}$$

$$A_A(t) = V_V + \tfrac{1}{2} t S_V.$$

Of course, for $t = 0$ the formulas (7.6) and (7.7) reduce to (7.1). If thick sections with at least two different thicknesses are available, Equations (7.6) or (7.7) can be solved for N_V, K_V, S_V, and V_V and, hence, unbiased estimators for these quantities result. In particular, this allows an estimation of the intensity N_V (as was first observed by Matheron (1976)) which is not possible by sections of zero thickness. If more variation of the thickness is possible, regression methods can be applied (see Voss and Stoyan (1985)). By combination of (7.6) or (7.7) with (7.1) simple approximation formulas for V_V (or S_V) can be obtained (see Stoyan (1985a) for details).

If X is a fibre process, then (5.16) (or (7.6) and (7.7)) gives

$$L_L(t) = \frac{\pi}{2} t^2 L_V \quad \text{and} \quad L_A(t) = \frac{\pi}{2} t L_V.$$

If we write $\tilde{N}_A$, respectively $\tfrac{1}{2} \tilde{P}_L$, for the intensity $\tilde{\gamma}_{Z_0}$, (5.18) gives

$$\tilde{P}_L(t) = \pi t L_V + 2\pi t^2 N_V \quad \text{and} \quad \tilde{N}_A(t) = \tfrac{1}{2} L_V + 2t N_V.$$

For nonisotropic processes X, the formulas are more intricate. We mention only the results for fibre processes. Of course, now the densities for the projected thick sections depend on t and on the direction x of the cylinder Z_0, $x \in \Omega$. From (5.15) we then get

$$L_L(t, x) = \pi t^2 L_V \int_\Omega |\cos \alpha(u, x)| \, d\tilde{P}_0(u)$$

and

$$L_A(t, x) = 2tL_V \int_\Omega |\sin \alpha(u, x)| \, d\tilde{P}_0(u),$$

(5.17) implies

$$\tilde{P}_L(t, x) = 2\pi t^2 N_V + 4tL_V \int_\Omega |\sin \alpha(u, x)| \, d\tilde{P}_0(u),$$

$$\tilde{N}_A(t, x) = 2tN_V + L_V \int_\Omega |\cos \alpha(u, x)| \, d\tilde{P}_0(u).$$

8. Final Remarks

The main purpose of this article was to transform general integral formulas for curvature measures into density formulas for geometric point processes. To this end, the integral geometric results, the fundamentals of geometric point processes, and the resulting general density formulas have been presented with the corresponding background and complete proofs, if necessary. The generalizations in Section 6 and the applications in Section 7 are treated in a much more cursory fashion, they have been included to show the variety and significance of the theory as a unifying approach to the many existing results in the literature. In some cases, the situations have to be studied in more details, and this will be done elsewhere. In others, the interested reader should be able to fill in the gaps.

More information about geometric point processes (in particular, higher-order properties which we have omitted completely), and applications in image analysis and stereology can be obtained from the book of Stoyan and Mecke (1983) (see also the forthcoming English version by Stoyan, Kendall, and Mecke). There one also finds additional literature, since our list of references is concentrated on those articles which are directly connected with the material presented here.

References

Ambartzumian, R. V. (1974) 'Convex Polygons and Random Tessellations', in E. F. Harding and D. G. Kendall (eds.), *Stochastic Geometry*, Wiley, London, pp. 176–191.

Ambartzumian, R. V. (1977) 'Stochastic Geometry from the Standpoint of Integral Geometry', *Adv. Appl. Prob.* **9**, 792–823.

Ambartzumian, R. V. (1982) *Combinatorial Integral Geometry*, Wiley, Chichester.

Cowan, R. (1980) 'Properties of Ergodic Random Mosaic Processes', *Math. Nachr.* **97**, 89–102.

Davy, P. (1976) 'Projected Thick Sections through Multidimensional Particle Aggregates', *J. Appl. Prob.* **13**, 714–722. Correction: *J. Appl. Prob.* **15**, 456 (1978).

Davy, P. (1978) 'Stereology – a Statistical Viewpoint', Thesis, Australian National Univ., Canberra.

DeHoff, R. T. (1978), 'Stereological Uses of the Area Tangent Count', in R. E. Miles and J. Serra (eds), *Geometrical Probability and Biological Structures*, Lect. Notes Biomath. **23**, Springer, Berlin, pp. 99–113.

Fava, N. A. and Santaló, L. A. (1978) 'Plate and Line Segment Processes', *J. Appl. Prob.* **15**, 494–501.

Fava, N. A. and Santaló, L. A. (1979) 'Random Processes of Manifolds in $\mathbf{R}^n$', *Z. Wahrscheinlichkeitstheorie verw. Gebiete* **50**, 85–96.

Goodey, P. and Weil, W. (1986) 'Translative Integral Formulas for Convex Bodies', *Aequationes Mathematicae* (to appear).

Gundersen, H. J. (1978) 'Estimators of the Number of Objects per Area Unbiased by Edge Effects', *Microscopia Acta* **81**, 107–117.

Jensen, E. B. and Sundberg, R. (1985) 'On Edge Effect in Planar Sampling', *Acta Stereologica* **4**, 89.

Jensen, E. B. and Sundberg, R. (1986) 'Generalized Associated Point Methods for Sampling Planar Objects', *J. Microscopy* **144**, 55–70.

Kellerer, A. M. (1985) 'Counting Figures in Planar Random Configurations', *J. Appl. Prob.* **22**, 68–81.

Kellerer, H. G. (1984) 'Minkowski Functionals of Poisson Processes', *Z. Wahrscheinlichkeitstheorie verw. Gebiete* **67**, 63–84.

Matheron, G. (1975) *Random Sets and Integral Geometry*, Wiley, New York.

Matheron, G. (1976) 'La formule de Crofton pour les sections épaisses', *J. Appl. Prob.* **13**, 707–713.

McMullen, P. and Schneider, R. (1983) 'Valuations on Convex Bodies', in P. Gruber and J. M. Wills (eds), *Convexity and Its Applications*, Birkhäuser, Basle, pp. 170–247.

Mecke, J. (1980) 'Palm Methods for Stationary Random Mosaics', in R. V. Ambartzumian (ed.), *Combinatorial Principles in Stochastic Geometry*, Armenian Academy of Sciences, Yerevan, pp. 124–132.

Mecke, J. (1981a) 'Stereological Formulas for Manifold Processes', *Probab. Math. Statist.* **2**, 31–35.

Mecke, J. (1981b) 'Formulas for Stationary Planar Fibre Processes III – Intersections with Fibre Systems', *Math. Operationsforsch. Statist., Ser. Statist.* **12**, 201–210.

Mecke, J. (1984) 'Parametric Representation of Mean Values for Stationary Random Mosaics', *Math. Operationsforsch. Statist., Ser. Statist.* **15**, 437–442.

Mecke, J. and Nagel, W. (1980) 'Stationäre räumliche Faserprozesse und ihre Schnittzahlrosen', *Elektron. Informationsverarb. Kybernet.* **16**, 475–483.

Mecke, J. and Stoyan, D. (1980a) 'Formulas for Stationary Planar Fibre Processes I – General Theory', *Math. Operationsforsch. Statist., Ser. Statist.* **11**, 267–279.

Mecke, J. and Stoyan, D. (1980b) 'Stereological Problems for Spherical Particles', *Math. Nachr.* **96**, 311–317.

Miles, R. E. (1974) 'On the Elimination of Edge Effects in Planar Sampling' in E. F. Harding and D. G. Kendall (eds.), *Stochastic Geometry*, Wiley, London, pp. 228–247.

Miles, R. E. (1978) 'The Importance of Proper Model Specification in Stereology', in R. E. Miles and J. Serra (eds.), *Geometrical Probability and Biological Structures*, Lect. Notes Biomath. **23**, Springer, Berlin, pp. 115–136.

Nagel, W. (1983) 'Dünne Schnitte von stationären räumlichen Faserprozessen', *Math. Operationsforsch. Statist., Ser. Statist.* **14**, 569–576.

Neveu, J. (1977) *Processus Ponctuels*, Lect. Notes Math. **598**, Springer, Berlin.

Nguyen, X. X. and Zessin, H. (1979) 'Ergodic Theorems for Spatial Processes', *Z. Wahrscheinlichkeitstheorie verw. Gebiete* **48**, 133–158.

Ohser, J. (1981) 'A Remark on the Estimation of the Rose of Directions of Fibre Processes', *Math. Operationsforsch. Statist., Ser. Statist.* **12**, 581–585.

Pohlmann, S., Mecke, J. and Stoyan, D. (1981) 'Stereological Formulas for Stationary Surface Processes', *Math. Operationsforsch. Statist., Ser. Statist.* **12**, 429–440.

Radecke, W. (1980) 'Some Mean Value Relations on Stationary Random Mosaics in the Space', *Math. Nachr.* **97**, 203–210.

Santaló, L. A. (1976) *Integral Geometry and Geometric Probability*, Addison-Wesley, Reading, Mass.

Schneider, R. (1979) 'Boundary Structure and Curvature of Convex Bodies', in J. Tölke and J. M. Wills (eds.), *Contributions to Geometry*, Birkhäuser, Basle, pp. 13–59.

Schneider, R. (1980a) 'Parallelmengen mit Vielfachheit und Steiner-Formeln', *Geometriae Dedicata* **9**, 111–127.

Schneider, R. (1980b) 'Curvature Measures and Integral Geometry of Convex Bodies', *Rend. Sem. Mat. Univers. Politecn. Torino* **38**, 79–98.

Schneider, R. (1981) 'Crofton's Formula Generalized to Projected Thick Sections', *Rend. Circ. Math. Palermo* (2) **30**, 157–160.

Schneider, R. and Weil, W. (1983) 'Zonoids and Related Topics', in P. Gruber and J. M. Wills (eds),
Convexity and Its Applications, Birkhäuser, Basle, pp. 296–317.

Schneider, R. and Weil, W. (1986) 'Translative and Kinematic Integral Formulae for Curvature
Measures', *Math. Nachr.* **129**, 67–80.

Schwandtke, A., Ohser, J. and Stoyan, D. (1986) 'Improved Estimation in Planar Sampling', *Acta
Stereol.* **6** (to appear).

Stoyan, D. (1979) 'Proofs of Some Fundamental Formulas of Stereology for Non-Poisson Grain
Models', *Math. Operationsforsch. Statist., Ser. Optimization* **10**, 575–583.

Stoyan, D. (1982) 'Stereological Formulae for Size Distributions via Marked Point Processes', *Prob.
Math. Statist.* **2**, 161–166.

Stoyan, D. (1984) 'Further Stereological Formulae for Spatial Fibre Processes', *Math. Operationsforsch. Statist., Ser. Statist.* **15**, 421–428.

Stoyan, D. (1985a) 'Estimating the Volume Density from Thin Sections', *Biom. J.* **27**, 427–430.

Stoyan, D. (1985b) 'Stereological Determination of Orientations, Second-Order Quantities and
Correlations for Random Spatial Fibre Systems', *Biom. J.* **27**, 411–425.

Stoyan, D., and Mecke, J. (1983) *Stochastische Geometrie*, Akademie-Verlag, Berlin.

Stoyan, D. Mecke, J. and Pohlmann, S. (1980) 'Formulas for Stationary Planar Fibre Processes II –
Partially Oriented Fibre Systems', *Math. Operationsforsch. Statist., Ser. Statist.* **11**, 281–286.

Voss, K. and Stoyan, D. (1985) 'On the Stereological Estimation of Numerical Density of Particle
Systems by an Object Counting Method', *Biom. J.* **27**, 919–924.

Weil, W. (1982) 'Inner Contact Probabilities for Convex Bodies', *Adv. Appl. Prob.* **14**, 582–599.

Weil, W. (1983a) 'Stereology – a Survey for Geometers', in P. Gruber and J. M. Wills (eds.),
Convexity and Its Applications, Birkhäuser, Basle, pp. 360–412.

Weil, W. (1983b) 'Stereological Results for Curvature Measures', *Bull. Int. Statist. Inst.* **50**, 872–883.

Weil, W. (1984) 'Densities of Quermassintegrals for Stationary Random Sets', in R. V. Ambartzumian
and W. Weil (eds.), *Stochastic Geometry, Geometric Statistics, Stereology*, Teubner, Leipzig, pp.
233–247.

Weil, W. and Wieacker, J. A. (1984) 'Densities for Stationary Random Sets and Point Processes',
Adv. Appl. Prob. **16**, 324–346.

Weil, W. and Wieacker, J. A. (1986) 'A Representation Theorem for Random Sets', *Prob. Math.
Statist.* **9** (to appear).

Wieacker, J. A. (1982) 'Translative stochastische Geometrie der konvexen Körper', Thesis, Albert-
Ludwigs-Universität, Freiburg.

Weiss, V. and Zähle, M. (1986) 'Geometric Measures for Random Curved Mosaics of $\mathbf{R}^d$', Preprint,
Friedrich-Schiller-Universität, Jena.

Zähle, M. (1982) 'Random Processes of Hausdorf Rectifiable Closed Sets', *Math. Nachr.* **108**, 49–72.

Zähle, M. (1984) 'Thick Section Stereology for Random Fibres', *Math. Operationsforsch. Statist., Ser.
Statist.* **15**, 429–435.

Zähle, M. (1986) 'Curvature Measures and Random Sets II', *Z. Wahrscheinlichkeitstheorie verw.
Gebiete* **71**, 37–58.